The Owner-Builder EXPERIENCE

How to Design and Build Your Own Home

by Dennis Holloway and Maureen McIntyre

Illustrations by Dennis Holloway
(except where noted)

 Rodale Press, Emmaus, Pa.

Many thanks to the following persons for their support, encouragement and assistance: Antonia and Robert Holloway, Douglas Holloway, Louise Jakimiec, Bertha Neal, Walter Brown, William Muschenhiem, Robert Metcalf, Bess Holloway, Usharbudh Arya, Shahram Ghadimi-Navai, Adam, Daniel, and Lauren Holloway, Charles and Muriel McIntyre, and the students of the Colorado Owner-Builder Center. Others who helped us with content, advice, and support include: William Angell, J. Douglas Balcomb, Bion Howard, Arnie and Maria Valdez, Mike Belshaw, Cheryl Holst, Bruce Gandrud, and the staff of the Boulder, Colorado, County Building Department. Sincere appreciation to the editors at Rodale Press for their advice and thoughtful attention to the final editing.

Printed in the United States of America on recycled paper containing a high percentage of de-inked fiber.

Senior Editor: Ray Wolf
Copy Editor: Cristina Negrón Whyte
Book Designer: Sandy Freeman

Library of Congress Cataloging in Publication Data

Holloway, Dennis.
    The owner-builder experience.

    Includes index.
    1. House construction—Amateurs' manuals.
I. McIntyre, Maureen.   II. Title.
TH4815.H65   1986   690′.83   86-13932
ISBN 0-87857-642-8   hardcover
ISBN 0-87857-643-6   paperback

2   4   6   8   10   9   7   5   3   1   hardcover
2   4   6   8   10   9   7   5   3   1   paperback

Contents

CHAPTER 1

The Owner-Builders

Everyone knows about owner-builders—those brave folks who design and build their own homes. We admire them because they have avoided mediocre conventional housing to produce unique living environments suited to their individual needs. They express an urge each of us has had since early childhood. Anyone who as a child rummaged for materials to build hideout huts, tree houses, or earthen dugouts will appreciate the excitement of the owner-builder experience. While the grown-up version of this experience may be more rigorous, the rewards of emotional satisfaction and pride that come with designing and building your own "place" are akin to the childhood experience. In fact, for most of us there are few undertakings in life that offer as much potential for personal growth, financial gain, and down-to-earth practical advantage as does designing and building your own home.

Building a house is not for everyone, but any man or woman who has the motivation and the persistence to get the required education—and who will take the time to properly organize the project—can be an owner-builder. If you have ever doodled an idea for a dream house, or if you are in the market for a new home but cannot find the "right" house on the market, then give owner-building a closer look.

Shelter is a basic human need, and each of us, if we stop and think about it, has unique ideas about what makes us feel sheltered and at home. Unfortunately, much of the housing available today is designed and built by people who have little more than a marketing or financial interest in the homes they produce. This may be one reason why owner-building has become so widespread. According to the Census Bureau, 22 percent of all new single-family American homes in 1982 were built wholly or in part by their owners.

WHY BUILD YOUR OWN HOME?

We are housing professionals. One of us is a licensed architect and the other is the director of the Colorado Owner-Builder Center. When we started working with owner-builders, we assumed they were motivated primarily by the economics of the housing market. Because owner-builders can save 20 percent or more by acting as their own contractors, and because the average cost of a new home is increasing so rapidly, we thought most of our clients were people who couldn't afford a home any other way. We were wrong. Most owner-builders take the project on because they want to, not because they have to. Their reasons for becoming owner-builders are instructive, and are echoed in the results of a recent nationwide survey of owner-builders conducted by *Rodale's New Shelter* magazine (now titled *Rodale's Practical Homeowner*).

Only 24 percent of the survey respondents cited cost savings as their primary motivation for building a house, although most did save significant amounts of money, with an average savings of 38 percent compared to purchasing a similar house. More than 26 percent of the respondents reported that the primary reason they decided to build for themselves was that they had always wanted to live in a house they had personally designed and built.

Many owner-builders build innovative, unconventional, and energy-efficient homes. Only 43 percent of those surveyed built homes with conventionally framed stud walls. Nearly one-third (32 percent) built timber frames, 11 percent built log homes, 4 percent built domes, and 3 percent constructed earth-sheltered designs. Respondents also reported that enviable amenities were incorporated into their homes—41 percent had workshops, 25 percent sport stained glass, and 23 percent contain saunas, hot tubs, or Jacuzzis. These homes reflect a concern for comfort and energy efficiency that is seldom realized in commercially built homes.

Because owner-builders are free to choose the particular technologies to be employed in their houses, they can combine from broad palettes traditional building techniques and state-of-the-art technologies to create homes that are economical to build, comfortable to live in, and inexpensive to heat and cool. For example, timber framing can be combined with stress-skin insulating panels, blending old and new technologies into a system particularly well suited to owner-builders (see chapter 4). We don't mean to imply that all owner-builders use unconventional building techniques. Conventional materials can also serve your needs well. The point is simply that as owner-builder, the choice is yours.

Another big plus to owner-building is the control that you exercise over the quality of the home. If you are the boss, you can see to it that things get done correctly. Eighteen percent of the survey respondents said they became owner-builders for this reason: They believed their houses would be of higher quality if they built the houses themselves. Owner-built homes tend to be less costly than comparable conventionally built homes, so owner-builders can use the money they save to buy higher-quality materials, such as ceramic tile, hardwood floors, solid-wood paneling and trim, and top-quality fixtures and hardware.

Owner-builders are an industrious group. The majority (69 percent) of the survey respondents were employed either full- or part-time while they built their homes. Most of those (55 percent), were employed full-time, and nearly half said their spouses also held full-time jobs during construction. They literally built their homes in borrowed moments during the week, on weekends, and during vacations! They are also a pragmatic group. Most used wood for heat, either as their primary source (62 percent) or as backup (37 percent), or they used some form of solar heat (65 percent). Their houses are well insulated, with nearly a third rating their walls at R-22 or better and ceilings above R-35.

Financing can be difficult for owner-builders for a number of reasons, which we'll discuss in chapter 6. Given a choice between lending to a contractor, with a track record of bringing projects in on time and budget, or to a novice owner-builder, lenders are apt to choose the contractor. This doesn't stop owner-builders from building their homes—they just figure out other ways to pay for it. Friends and family will often make loans with more competitive rates and more confidence in your abilities than will a banker. Only 34 percent of the survey respondents used financing from lending institutions, while most relied primarily on their own savings.

Based on the survey and our own experience, we believe a profile of an average owner-builder would look something like this: A white married male in his thirties who has never built a house before, who makes between $25,000 and $30,000 a year, and who already owns a home. But this profile can be misleading. Our experience is that owner-builders represent all facets of society, including single women, grandparents, and low-income families. Because these people are highly motivated, they find ways around any obstacles that arise. Twenty-one percent of those surveyed claimed that they encountered no major problems at all, while others listed the high cost of materials, their low basic construction skill level, and the difficulty of getting financing as obstacles. Only 20 percent considered any of their problems to be major, except perhaps for the high cost of materials.

It should be clear by now that owner-builders are a resourceful group. Fully 55 percent built the foundation themselves, 83 percent did their own framing, 82 percent laid the floors, 76 percent installed their own doors and windows, 73 percent did their own roofing, 55 percent did their own electrical work, and 53 percent plumbed the house themselves. Surprisingly, 89 percent of the survey respondents did not get professional help to draw their designs.

Despite their high degree of self-reliance, most of the respondents got at least some help with their homebuilding projects. Even those diehards who dug their own excavations and mixed their own concrete called in family and friends at some point to help raise walls or frame the roof. Of the owner-builders surveyed, 74 percent hired help to put in the driveway, 57 percent hired a surveyor, 54 percent bought cabinets or had someone build them, and 52 percent hired out the excavation work. In fact, some of the owner-builders never picked up a tool at all, choosing instead to act as general contractors—coordinating the delivery of materials and scheduling the subcontractors who did the actual construction work. We will discuss this route more in chapter 2.

SOME ADVICE

Although most owner-builders we know consider their housebuilding project to be a major, positive event in their lives, they also admit that it did place a great deal of strain on them and their families. Most are free with advice about things they wish they'd known when they started and about difficulties that could have been avoided with more planning.

First of all, no matter how long you think it's going to take to build your house, it will take longer. Many owner-builders wish they had spent more time on planning and design. The amount of money you'll save on your project and the ease with which you'll accomplish it are more dependent on planning than any other factor. The time you spend designing and redesigning, exploring alternatives, estimating costs, shopping for "deals," and studying other owner-builders' experiences is the most valuable time you can spend. We find again and again that the people who ran into major problems and made expensive mistakes are the ones who didn't do the necessary planning or didn't want to spend the money to take a housebuilding class or consult with professionals.

It is always easier to rearrange walls, rooms, doors, or windows by erasing and redrawing them on paper than by moving them physically after the house has been framed. The only way we know of to avoid big, costly mistakes is to repeatedly think through the entire project until you feel like you could build the house in your sleep. Still, as construction proceeds, almost every builder discovers some details that require rethinking. The idea is to do enough preplanning so that these changes will be as small as possible. Because your project will undoubtedly cost more than you initially think, it's wise to pad your budget with a safety margin. Try to plan positive surprises into your work.

If at all possible, postpone moving into the house until it's finished. Living amid sawdust and lumber scraps adds substantially to the strain on individuals and families. When there isn't an uncluttered spot in the house in which to relax for a while or to get away from the rest of the family, tempers can flare. Of course, since it may be difficult to maintain your existing residence, you may be forced to move into your unfinished new home prematurely. If this happens, finish one or two rooms to use as living quarters while working on the rest of the house. And try to budget enough money to treat yourselves periodically to a weekend at your favorite "second honeymoon" spot. Of the survey respondents, 62 percent lived elsewhere while building, 25 percent lived in a temporary structure or trailer on site, and 21 percent succumbed to the temptation to move in before the house was done.

Finally, consider carefully how much of the work you can do yourself. It is generally considered good economics to do those jobs that (1) take the most labor, time, or money, but (2) interfere the least with your subcontractors' work. This is not a hard and fast rule, however. There are jobs that we suggest you leave to professionals. For example, a mistake on a concrete pour can be costly and demoralizing at the outset of your project. We recommend hiring a pro to inspect the forms and help with the pour, and hiring a finisher for the concrete "flatwork" if you're inexperienced in such work.

GETTING GOOD INFORMATION

We have done our best to present in this book the most up-to-date information on the various aspects of owner-building. Owner-builder products, publications, and organizations are more numerous than ever before. One of the best places to go for more information is your local owner-builder organization or school (see Appendix B). The schools offer varying services, but they all share an orientation toward energy-efficient, intelligent housing that meets the needs of the occupants. Most offer classes, seminars, and workshops in all phases of the building process, as well as consultants who can help you work through the snags that you run into during the project. Many also act as informal "networking" centers, putting owner-builders in touch with each other, guiding them to useful government agencies and receptive lenders, and helping them find well-qualified architects, realtors, tradespeople, and other sources of expertise and information.

The first such school, Shelter Institute, was founded in 1974 in Bath, Maine. The largest school currently is the Owner-Builder Center in Berkeley, California. The Center gave the owner-builder movement a boost by training people interested in starting their own centers, and there are now over 30 such groups operating around the United States and Canada.

One cautionary note on working with owner-builder schools, especially those recently established: Before you pay your money to register for courses, check the credentials of the teachers, since there is no licensing procedure for teachers at owner-builder schools. It is up to you to verify that the teachers, like anyone else you turn to for advice and information, are competent and well grounded in real-world experience. Fortunately, our own experience is that employees at owner-builder schools tend to be highly motivated, as well as highly skilled. Still, it would be worth your while to talk with former students to get their evaluations of particular courses and teachers.

OUR BIASES

We bring certain biases to the writing of this book. We share them with you here to help you evaluate the information we present in the rest of the book. By knowing our underlying assumptions, you will be in a better position to decide whether you agree with our specific recommendations.

Building a house for yourself may seem like a new and revolutionary concept, but only since modern times have most people not built their own houses. To some, this is liberation, just as not having to work the land for food seems to be liberation. But we believe such liberation is illusory: Giving control of the design and construction of our personal living environments to an impersonal housing industry is actually a loss of control, a real loss of freedom.

Personal participation in determining the form of the houses we inhabit is more important than most people understand. The interplay between people and their living environments is a classic chicken-and-egg story. We design and erect buildings, shaping them as we see fit. But then we dwell in these buildings, and the quality of our lives is deeply influenced by them. In recent decades, we have given up control of our home environments to architects, contractors, developers, tradespeople, and bureaucrats who, however well intentioned, are too remote from us to understand our individual needs. When building standardized housing units, developers concentrate on speed and efficiency to maximize profits, and therefore they avoid involving the homeowners in the process, since this could cause delays.

We don't claim that the owner-building movement is a cure for all the world's problems, but it can help reestablish local self-reliance. Before that can happen, however, the network of owner-builders needs to be strengthened and broadened, and this is our challenge in the next decades. Owner-builders often approach the creation of a new home too individualistically. This may solve one family's personal needs, but may result in alienation from the local community. In the future, owner-builders need to consider the advantages of pooling their labor and time to construct communities of homes rather than isolated dwellings.

Beyond the neighborhood there exists the challenge of the larger environmental context—a world of finite resources dangerously overburdened by pollution. Our suggestions will always favor solutions that conserve energy in the home to minimize the necessity for burning fossil fuels. We also prefer renewable materials and products that do not pollute the indoor environment and that don't require massive amounts of energy to manufacture.

We recognize that your priorities may be different from ours, and we invite you to use the information we've provided in any way that proves useful to you. One of the inspirational things about working with owner-builders is the fresh perspective they bring to their housebuilding projects, so that often we, as teachers and consultants, become the students. In this book, we offer you the knowledge we have gained by working with owner-builders, and we encourage you to pass on what you learn in realizing your new home to the others who are sure to follow.

FIVE OWNER-BUILT HOMES

Kendall Home

Suzanne and Bob Kendall wanted a home that would last, so they built theirs with concrete block walls insulated on the outside with expanded polystyrene finished with stucco. From the front, the 4,600-square-foot home looks conventional, but its passive solar features are evident from the rear. (Photos by Thomas M. Eiben)

Pryce-Tarolli Home

Douglas Pryce, an attorney, took a year and a half off from his job to build this passive solar home for himself and his partner, Mia Tarolli. He spent several years designing the house, which incorporates a two-story sunspace. The house—located in the mountains west of Denver—was being finished when these photos were shot. (Photos by Thomas M. Eiben)

Valdez Home

Arnie and Maria Valdez designed and built their adobe home in the San Luis Valley of southern Colorado. Containing 3,000 square feet of living space, it cost about $10 per square foot to build, and—thanks largely to its passive solar design—it uses just one cord of firewood for heat and cooking each winter. (Photos by Mitchell T. Mandel)

Young Home

Carol Young spent two years researching before she began construction of her timber frame home. Then, during construction, she took a year off from her career as a ballet instructor to act as her own general contractor. To keep costs under control, she used 12" × 12" recycled timbers from the tipple house of an old mine. For energy efficiency, she made the walls with foam-core panels. (Photos by Maureen McIntyre)

Stricklin Home

David and Patti Stricklin hired architect Dennis Holloway to assist them in designing their three-story owner-built home. They worked on the house for three years. The strongest exterior features of the home— suggested by Holloway—are a solar tower and two flanking Trombe walls. (Photos by Thomas M. Eiben)

CHAPTER 2

So You Want to Build a House

Building a house is a complex and challenging undertaking. But, while it's not a job to be undertaken lightly, the skills and information needed are readily available to anyone willing to take the time and energy to track them down. Above all, don't rush into a housebuilding project. Allow plenty of time to do the research and planning necessary to assure that your home satisfies your needs. The most successful owner-builder projects we are aware of were years in the planning. As a general rule, 75 to 80 percent of the time you devote to your project should be spent *before* construction starts.

Planning a home is essentially a process of making thousands of decisions, from the very general ("Do I really want to build a house for myself?") to the very specific ("What color ceramic tile goes in the upstairs bathroom?"). These decisions, especially the early ones, will determine how well the project goes. For most owner-builders, the decision-making process settles into a series of compromises with family members, lenders, building officials, subcontractors, and suppliers. Practical considerations (such as the resale value of the finished home, budget constraints, and so forth) also enter the picture, requiring further compromises.

Professional builders rely heavily on their past experiences to plan and manage construction jobs. Since owner-builders generally have no previous experience to fall back on, they must spend much more time on the project. One way to reduce this load is to delegate various tasks to professional contractors. Indeed, deciding how much of the project you really want to do yourself is one of the most crucial decisions you should make. You could decide to do it all, or you could decide to farm out many jobs.

LEVELS OF INVOLVEMENT

The minimum level of owner-builder involvement is acting as your own contractor. If you elect this option, you will not perform much of the actual construction work, but you will oversee the project. Sounds easy. But even this "minimum" involvement involves broad responsibilities: You will plan all the stages of the project, hire workers to perform them, coordinate the schedules of the various workers, arrange for the needed materials and supplies, and ensure that all the work meets the applicable codes and standards. It's a big job, but it can be financially rewarding. You may save 20 percent compared to the cost of hiring a general contractor to build a house for you.

Be Your Own Contractor

To help you decide whether you feel able to operate as your own contractor, here's a more detailed description of how a professional general contractor operates. A "general" orchestrates the materials suppliers, subcontractors (contractors who work under the general contractor, such as carpenters, electricians, etc.), and equipment operators (bulldozer operators, etc.). The general hires and supervises these individuals. At the same time, he or she deals with the architect (if there is one), building inspectors, the lender, and the owner.

The general often arranges the financing, since he or she probably has a relationship with a lender who makes construction loans. Because the lender is familiar with the general contractor's work and has worked with this individual successfully on past projects, the process of getting the loan is smoother. A general who has been in business for a while also has an ongoing relationship with the building department and other local governmental agencies, such as zoning and planning departments that can influence building projects.

The general is first and foremost a manager, someone who schedules deliveries of materials and assures that the appropriate workers will be on site to install these materials. The general is also responsible for making sure there are no gaps in the project—that each task is completed adequately so that the next subcontractor can take over (for instance, the general makes sure that the carpenters prepare the bathrooms adequately for the plumbers to be able to do their work). He or she also schedules so that tradespeople don't interfere with each other. Once the house is framed, many things are happening simultaneously, and careful scheduling will keep the subcontractors working efficiently.

When problems arise, it is the general who is consulted, and usually it is the general who resolves them. These problems can range from lack of availability of a specified material to settling disputes between tradespeople. (If you hire an architect to provide "comprehensive" services—see chapter 3—the architect can be counted on to help resolve these problems. In that case, the subcontractors' contracts should identify the architect as the arbiter of problems.)

Finishing Out a Shell

Whether or not you decide to function as your own general contractor, there are several other roles you may either fill yourself or farm out. For example, one popular option among owner-builders is to have the foundation poured and the house framed by professional workers, then finish the interior of the house as a do-it-yourself project. A home-construction project feels much more manageable if the "shell" of the house is already up when you begin your share of the work. According to the Owner Builder Center in Berkeley, California, you can reasonably expect to save 40 to 44 percent of the total cost of a contractor-built home by taking this route.

There are a number of "kit" home companies that specialize in helping owner-builders do this. They provide the basic structure of the house in the form of a kit that their own workers or local contractors will assemble for you. Then you can perform the finishing work. In choosing a kit-home manufacturer, look for one that will be available for technical

(continued on page 16)

Table 2-1 WORK BREAKDOWN

Job	Budget Estimate ($)	Crew Days	Time Periods: (fill in date)								
			1	2	3	4	5	6	7	8	
Excavation											
Concrete											
Masonry											
Framing carpentry											
Insulation											
Roofing/ Gutters											
Plumbing											
HVAC											
Electrical											
Finish carpentry											
Stair construction											
Drywall											
Exterior painting											
Interior painting											
Ceramic tile											
Floor covering											
Landscape											
Total											

	9	10	11	12	13	14	15	16	17	18	19	20	21	22	23	24

support during your portion of the project—if you run into a problem, the company should willingly give you advice and guidance. The materials provided in the kit should be high quality, and the kit package should be complete. Always talk with former customers about their experiences with the company and visit their completed homes. If the salespeople seem vague or evasive, shop elsewhere. (We will discuss kit homes in more detail later.)

Deciding to "finish a shell" does not necessarily mean you must do all the finishing work yourself. You can elect to hire subcontractors to perform particular tasks. Which jobs you decide to do yourself is a decision only you can make, but there are some parts of the project that, in our view, make more sense to do for yourself than others. We encourage you to do your own insulation and weatherproofing. This is not particularly skilled work, but it takes a degree of care and attention to detail that most subcontractors are unable to provide because of time and economic constraints. Your "sweat equity" will be repaid many times over in reduced energy costs.

Plumbing and electrical work are relatively light and enjoyable, and a great deal of money can be saved since plumbers and electricians often charge dearly for their services. Make sure that local codes allow you to do these types of work, however. In fact, no matter what types of work you want to do yourself, you should check to see whether local ordinances allow you to do so. Some building departments restrict what work they will allow "amateurs" to undertake.

Ceramic tile installation is skilled work, but it is easy to learn, and novices can substitute time and attention for expertise. The work is enormously satisfying, because the results are so attractive, and ceramic tile adds value to the home. A tile job is only as good as the adhesive and backing material used, so get professional advice about what works best in your area.

If you plan to do your own finish carpentry but have little or no carpentry experience, it's a good idea to gain experience by "apprenticing" yourself to the professional carpenters who erect your house's shell. This will give you an opportunity to work with wood and power tools without worrying about how the finished product will look—the work on the frame will be hidden inside the finished surfaces of the house. Finish carpentry requires care and precision, but a novice can allow more time than most professionals, so the result can be as good, and sometimes better, than a professional job. If you don't own the necessary tools, they can be rented reasonably in most areas. A power nailer, for instance, makes short work of laying flooring, and a power miter saw will speed the installation of solid wood paneling and trim.

There are some tasks we think you should *not* do yourself. One of these is drywall finishing work. Professionals will do a much better job than you could in a fraction of the time. If you do the job yourself and make mistakes, you'll see those mistakes every day for years afterward. Some professionals insist on hanging as well as taping and finishing the drywall, so check before you hang it yourself.

If your roof is complicated or excessively steep, hire a roofer. Most professional roofers warrant their work—if there's a leak during the warranty period they will repair the roof for free. Although the mechanics of installing a roof are straightforward, mistakes that result in leaks are annoying and expensive to repair. Make sure the flashing system is designed and installed properly—inadequate flashing can cause no end of grief.

Be sure to develop contingency plans, in case the unexpected occurs. You may be prevented from doing all the work you had planned to do, so you may need to hire workers to take up the slack. You should arrange your budget so that you will have sufficient funds for this purpose. One tangential benefit will be that, if you are financing the construction through a bank, your banker will be more comfortable with the project—the bank's loan will seem that much more secure, since you will be prepared to finish the house no matter what.

Doing It "All" Yourself

Although most owner-builders get help with part of the project, there are those hearty few who do virtually all of the physical work themselves. It is possible to save 50 percent or more of the total cost of a comparable contractor-built house this way, but it's impractical unless you can take a year or so off from your job. Charlie Wing, founder of Cornerstones Energy Group, Inc., an owner-builder school in Maine, estimates that it takes about 1½ hours per square foot to complete the average home. Thus, building a 1,500-square-foot home would take about 2,250 hours or roughly 47 weeks.

Ideally, you should finance the house out of pocket so you don't have a lender hounding you to finish the work quickly. But even in this ideal arrangement, you need to plan for the timely completion of the project. Make a tentative schedule so that you can monitor your progress and do a detailed cost estimate to help you think the project through. It is easy to lose perspective when you spend day after day working alone on such a large project. You can use your schedule and cost estimate to assure that you follow a logical building sequence and to avoid getting frustrated by what can sometimes seem like an unending process.

Single-handed work will go more easily if you select your construction materials wisely. Various materials are made to order for do-it-yourself builders. Surface-bonded block and interlocking block systems take relatively little skill to assemble, and the end product is a stronger wall than a mortar-jointed concrete block wall. Prehung doors are much easier and quicker to install than conventional doors, and they come in a wide variety of styles and materials, including expertly weatherstripped exterior doors. (Chapters 4 and 5 give further information about building systems and materials.)

You can save money by using salvaged materials. If a neighbor is tearing down a garage, for example, you might volunteer to help in exchange for the materials. It is also possible to obtain free or very inexpensive materials, such as seconds, damaged freight, or leftovers from suppliers and manufacturers. You can turn the fact that we live in a throwaway society to your advantage—but you must be prepared to pay a penalty in terms of time. Searching for the materials will take time, and using them will probably call for careful cutting and fitting.

ASSESSING YOURSELF

Ultimately, your decision about how much work to undertake yourself must hinge on your assessment of your own skills and strengths. To help uncover the talents and strengths you bring to your housebuilding project, as well as the practical constraints you'll be working within, here is a list of questions you and your family might ask yourselves. We'll start with a seemingly objective—but touchy—subject: money. Then we'll proceed to more subjective matters.

Money

Can you afford to build your own home? Do you *want* to spend your money for this purpose? Mull over the following questions:

■ How much does housing cost in your area? To get some idea of what you'll need to spend on your new home, find out what homes are selling for and how the costs break down. For instance, is land at a premium? Are there restrictive codes, covenants, or zoning restrictions that are driving up costs? Are labor costs particularly high?

Remember that the home you build is likely to be unique, and you won't benefit from the economy of scale achieved by tract builders who build many identical homes. You should learn the selling price of customized homes that are comparable in quality to the home you want to build, and then adjust this amount downward by approximately 50 percent if you will do virtually all the work yourself, 40 percent if you will finish out a shell, or 20 percent if you will act as an owner-contractor. This will give you a rough idea of your home's potential cost.

■ How much can you afford to pay for housing? It's a useful exercise to sit down and think about how much you're accustomed to paying for living space, and whether you are comfortable with that. You should also consider how you will fund the project.

If you aren't sure how large a house you can afford, visit mortgage lenders and ask them to qualify you for a mortgage. Most mortgage companies are willing to do this, anticipating your business. For instance, suppose you qualify for an $80,000 mortgage, and you already own your lot. The lot is worth $25,000, which gives you enough equity in the project to make the construction lender comfortable. That leaves the $80,000 for developing the lot and building the house. Assuming that contractor-built custom housing in your area goes for around $50 per square foot, and you plan to act as your own contractor (which should save you roughly 20 percent of the total cost), you could build up to a 1,900-square-foot house. These are preliminary figures, but they give you a place to start. If it turns out that you can qualify for only enough to build a 500-square-foot house, you don't own a piece of land, and you have no cash for a down payment, you would probably be well advised to start saving or explore the possibilities of borrowing from family and friends.

■ How much money are you *willing* to spend? For some people, how much they can *afford* isn't the issue. It is a point of honor with them to build a house that is functional and aesthetically pleasing for the least possible dollar amount. These are the folks with the tenacity to tear down old buildings for materials, dig their footing trenches by hand, and run every pipe and pound every nail in the structure. If you fall into this category, we applaud you and urge you to take the same care in designing and planning that you would if you were buying lumber at the lumberyard and borrowing money from a bank. It is impossible to emphasize too much the time and energy you will save by thinking the project through before you start to build.

Building Skills and Time

Besides making basic decisions about money, you need to decide who will do what in the construction project. The following questions should help you sort things out:

■ How much of the labor do you plan to do yourself? Realistically, how quick a worker are you? How many hours a week can you devote to the project? What activities will you be giving up to work on the house? If those activities usually include other people, especially family members, how will they feel about their reduced access to you for the duration of the building project?

■ How much do you know about housebuilding? We have a recurring experience at the Colorado Owner Builder Center of people coming in who know at least a little about housebuilding, and on that basis decide that they don't need the whole program and will just avail themselves of the files, library, etc. These are almost without exception the same people who call every couple of days with questions and problems. One man finally decided to take the housebuilding course and came to it with $4,000 worth of construction drawings, only to toss out the drawings halfway through the class when he realized the design was not at all what he wanted. He was actually much more cheerful about it than we were; he had been saved from building a house that would have cost a great deal of money and ultimately wouldn't have served his needs.

■ How much experience do you have with intricate work? If you like the sense of accomplishment that comes with taking on a fairly complicated project and completing it successfully, whether it be sewing a complex garment or rebuilding the engine of your car, chances are you'll enjoy your housebuilding experience. If you're the sort of person who is good with your hands and takes pleasure in working with tools and doing little projects around your house, you are owner-builder material.

■ How proficient are you with tools, particularly power tools? If your blood pressure rises every time you get near a spinning blade, you'll probably want to hire carpenters. People who work with power tools regularly walk a thin line between a healthy respect for the damage they can do and gratitude for the speed and accuracy such tools add to a task. Working under such conditions day after day requires a constant presence of mind.

■ If you have any experience with physical work, did you enjoy the experience? If you plan to do much of the work yourself and have little experience, we would urge you to loan yourself to a contractor to get an idea of what it's like to work physically for long hours day after day. You risk days of sore muscles and possible injury if you don't ease yourself into the regimen, and it would certainly pay to find out now if the joys of physical labor are for you. There are any number of ways you can be actively involved in your project without exerting yourself physically.

If you don't exercise regularly, consider developing an exercise regimen to increase your physical strength and endurance. If you are overweight, the building project might be an incentive to normalize your weight. If you are unaccustomed to exercise in any form, or if you are significantly overweight, talk with your doctor before starting any exercise program or reducing diet. Even people in good shape should take reasonable precautions. Temperatures on a roof in the middle of summer, for instance, can strain even the healthiest body. One woman started working out and swimming every day about six months before she started building her home; she swears it saved her, once the project began. Accidents typically happen when workers are tired or harried, and your physical and psychological stamina are at least in part a function of how healthy and fit your body is.

■ If you anticipate assistance from family and friends, how likely is it that the help will be forthcoming when you need it? Careful organization may help you to harness the energies of enthusiastic but unskilled volunteers. The spirit of cooperation and community that such successful group efforts foster is energizing and infectious, but the undertaking requires thoughtful planning on your part.

Decision Making

How are your managerial and organizational skills? As owner-contractor, you will manage the project. Although lack of contracting experience might seem like a serious liability, you probably have the required skills if you are a good manager. Housewives, for instance, are masterful schedulers, particularly if they have part-time jobs. And most parents have become adept at conflict resolution, a skill which is sometimes necessary at the building site.

Don't be intimidated by your lack of specific technical information. Most of what you need to know will be available from your subcontractors. For example, if you're not sure where the excavator's responsibility ends and the concrete contractor's begins, ask them both. Most subcontractors are happy to give you this kind of information, and often you can put them in touch with each other and let them work out who's going to do what. We urge you to stay involved in how they work it out and clarify in the contracts you have with both of them what their individual responsibilities are.

As is so often the case in any undertaking, the most valuable skill you can bring to your project is the ability to deal fairly and effectively with other people. The stickiest hassles owner-builders get into inevitably involve disputes with subcontractors, suppliers, building officials, lenders, and spouses. Approaching each day's work with a sense of humor and patience can go a long way in making the project enjoyable and efficient. People rarely work at their best when they feel pressured or abused. If you are respectful toward the people around you, you'll get better work from them, and you will all have a more positive building experience.

One owner-builder couple had it down to a science. They designed and built a beautiful passive solar home, largely by themselves, but with a great deal of help from their friends. The wife scheduled volunteer help the way most people schedule tradespeople, contacting the volunteers shortly before they were expected. She gave them instructions on what tools to bring, what job they would be working on, and the hours she expected them to be available (she had everyone working 4-hour shifts). She even told them what clothes to wear (heavy boots for framing, rubber boots for concrete work, rain gear if the weather looked threatening, etc.). She now proudly states that all the people who worked on the house are still friends, and the project came in on time and on budget. We should add that she was holding down a job and finishing a master's degree—and she was pregnant—while all this was going on.

■ What is your decision-making style? Do you make decisions quickly and without an excess of waffling and worrying? If you are building with someone else, whether a spouse or other partner, take a look at how well you make decisions together.

Some people avoid major conflicts by assigning areas of responsibility and establishing a set of ground rules that outline when the partner must be consulted. For instance, one partner could be in charge of color schemes, but would be obliged to consult the other before reaching any hard decisions. This decision-making strategy works well as long as the individuals affected by the decisions are mature enough to live with the results once the house is built. Reminding each other of less-than-perfect decisions made under stress is neither useful nor kind and won't contribute to harmonious relationships. Some owner-builders include in the ground rules an agreement that complaining about things that can't be changed will be kept to a minimum.

It is nothing short of miraculous how mistakes that seem like insurmountable obstacles shrink to remembered milestones by the end of the job. Somehow, the house gets completed, and the despair you felt staring at the bend in your new foundation wall fades into the total sense of pride at having built your own home.

■ Do you have the emotional stamina for housebuilding? There is something intensely personal about shaping one's own shelter, even if your function is coordinator/overseer rather than pounder of nails. Emotional flexibility is an enormous asset in stressful situations, and the ability to detach yourself periodically and take a fresh look around will help you enjoy the process when you come back to it. If you're doing the work yourself, many of your frustrations will likely be worked off pounding nails and lifting walls, but if you're supervising, get some exercise and find time to get away from the project.

Occasionally we run across clients who have decided to build a house in an attempt to cement an otherwise shaky marriage. Shared challenges and triumphs will serve to strengthen and enrich families who have learned to be resilient and giving, but they will almost always intensify misunderstandings for those who start the project with basic issues left unresolved. The divorce rate among owner-builders is distressingly high, and we suspect it is largely because couples hope that the mutual focus and involvement will somehow bring them closer and heal the rifts that have developed in their relationships. In many cases, the opposite turns out to be the case. The added pressure turns rifts into chasms, and communication breaks down altogether.

To guard against such problems, you should try to anticipate the strains in order to be patient and understanding with each other. Planning and gathering good information go a long way toward minimizing potential disputes. This is not to suggest that both partners must be intimately involved in the project for the house to be successfully completed. The ideal is to have everyone in the family enthusiastically involved, but many homes have been built by one partner while the other kept a regular job to provide a steady income. The rule here is to do whatever works for you and yours. Even if some family members don't want to be actively involved in construction, they can still be invaluable as sounding boards and sympathetic shoulders.

FINDING SUBCONTRACTORS

Finding reliable subcontractors is no easy job, especially for someone unfamiliar with the building process. Word of mouth is always the best advertising, so begin by asking around. If a poll of friends and relatives doesn't turn up anything, check at local materials suppliers. Suppliers are unlikely to recommend "subs" who don't pay their bills, so you will at least have some assurance that the tradesperson is financially responsible. If you are working with an architect or other design professional, he or she may know of tradespeople who have proven reliable. If there is an owner-builder school in your area, check to see if it has files of tradespeople interested in working with owner-builders.

Construction lenders might also be a source of contacts, and visits to construction sites to quiz general contractors and subcontractors has turned up leads for some novice builders. It might even be worth buying an hour from a general contractor or two to get the names of subcontractors they use. Don't be offended if the contractors are reticent about sharing this information with you. They may consider it proprietary, since reliable subcontractors are so valuable, and they may not want to risk their subcontractors being tied up on your job when they need them.

Pulling subcontractors blindly out of the yellow pages is not a good idea. Anybody can run an ad, and although most subcontractors are honest, hardworking people, consumer complaints arising out of relationships between homeowners and subcontractors are among the most frequent problems taken to consumer agencies and small-claims courts.

Evaluating Subcontractors

Regardless of how you find subcontractors, always ask for references, and check them. Ask the subs' former clients if they showed up when they said they would, if the job was completed on time, if they cleaned up after themselves, if there were any major problems as work progressed, if there were cost overruns, if the clients would hire the subs again. Do enough homework ahead of time so that you have some sense of what constitutes a "workmanlike" job. Then look at several completed jobs the subcontractors performed, and, if possible, visit the jobs they are currently working on. Much of what you can see at a completed job site is finish work—trim carpentry, drywall finish, etc.—which may hide various flaws, so your conversations with the clients are at least as valuable as your personal observations.

Good subcontractors seem to be always busy and can be hard to reach. It is worth the time and trouble to track them down. Listen carefully to their thoughts about your project. They do this kind of work every day and probably know how to get the job done less expensively and more quickly than you do. Subs have a different perspective on building than architects and building officials. They know what works best in the real world, and most are happy to share the information, if you have a receptive ear. They can give you suggestions on what materials to use, how long the job will take, and roughly how expensive it will be.

You can also get a feel for the aesthetic sense and general attitude of the subcontractors as you visit with them. Trust your intuition; if you immediately dislike or distrust some subs, don't hire them no matter how good a deal they're offering or how highly recommended they come. It's unlikely that your working relationship will be a good one if it starts on that sort of footing. You don't have to become fast friends, but you ought to have a generally positive feeling about your subs.

Let each subcontractor know when you need the job done, and see how it fits into his or her schedule. Don't take it personally if a subcontractor decides not to submit a bid. If a sub doesn't have the time for your project, you are better off learning this now than halfway through the job.

Subcontractors fall into roughly two groups: specialists and generalists. Specialists are more common in urban areas and typically do one part of a job exclusively (e.g., plumbing, electrical work, framing carpentry, or finish carpentry). Generalists are more likely to be found in rural areas and can often build a house from the ground up, including the mechanical systems. There are advantages, disadvantages, and appropriate uses for both types of professional.

■ Specialists have the advantage of an intimate familiarity with their particular trade, and most take real pride in keeping up-to-date with new developments and doing the best job possible. Specialists are often more expensive than a general builder in terms of their hourly rate, but they are also often faster, so the higher rate is offset somewhat. In some areas, trades are protected by unions or restrictive building codes, so you have no choice but to hire a licensed subcontractor. In other areas, you may do the work yourself, but if it is to be hired out, it must be done by a licensed professional.

■ A generalist has the advantage of familiarity with the entire house as a coherent whole, in contrast to a specialist's view from the relatively narrow perspective of his or her trade. Many owner-contractors who can't be on site every day hire builders in the capacity of construction foremen. Such foremen are valuable not only to keep an educated eye on things as the work progresses, but also because they can work along with the subcontractors. Hiring a generalist foreman can cut down the number of trades you have to subcontract out, which simplifies your scheduling somewhat.

Particularly in the case of carpenters, you'll want to be sure you have the right person for each job. When a house is being framed, walls are often "tapped" into place with that gentle persuader, the sledgehammer. Framers work quickly and usually needn't be concerned with the appearance of their work, since it will all be covered up. The skills and temperament necessary to do finish carpentry are very different. Such woodworking requires patience and careful attention to detail. Before you hire one person to do all the carpentry in your house, take a careful look at the finish work he or she has done and make sure the quality is up to your standards. You might also want to have this carpenter bid separately on the framing work, since if he or she takes the same care with framing as with finish work, you could end up with a perfect but very expensive framing job.

Soliciting Bids

Before you can ask subcontractors to bid on your job, you must have a set of plans and specifications (see chapter 3 for details on designing and chapter 5 for specifics on materials). A specifications list describes the materials that will be used to build the house. To protect themselves, subcontractors will probably inflate their bids somewhat if your specifications are unclear, so be as specific as possible. Many lenders have "Description of Materials" forms that can help you organize your thinking.

You may want to include in the specifications a statement that all materials will be installed according to manufacturers' directions. Steer clear of "or equal" clauses, since what seems equal to a sub may not seem equal to you. It's safer to require the subs to use the specific materials you name (spell out the brands, colors, and sizes you want).

Be specific about the parameters of the work you want done. For instance, how far outside the building foundation does the plumber's responsibility extend? If he includes only the cost of extending the water and sewer 5 feet beyond the house foundation, who's going to take it the rest of the way to the street? Go through your PERT charts (see Appendix A) and make sure that somebody has clear responsibility for each task. If there are gaps, you can often pay the subcontractors to complete the necessary tasks, or, if you have a generalist builder on the job as foreman, that person can sometimes fill in.

Get at least three bids for each part of the project. It takes considerable time and effort to work up an accurate bid, so if you ask subs to bid a part of the job you're planning to do yourself (to satisfy a lender, for instance), offer to pay them for their time. If you ever need their talents in the future, they will likely remember the courtesy.

Don't automatically accept the lowest bid. Your main concern should be the quality of work and the reliability of the contractor. There are any number of things that can drive up the cost of a job once it starts, and all too often the low bid ends up costing considerably more over the long run. Even on a fixed-price bid, it isn't in your interest to have the subcontractor losing money and feeling bitter about it, since you're not apt to get the quality of work you want under those conditions. If the low bid is much lower than the others, either throw it out or give the sub the opportunity to refigure it. Creating and maintaining a good working relationship with subcontractors is always worth the effort.

There are essentially two types of bids you're likely to encounter. The most common is the "fixed-price" bid. The subcontractor figures materials, labor, and profit, and gives you a firm dollar figure of what it will cost to complete the job. Some jobs are bid on a "cost-plus" or "time and materials" basis. In this case, the subcontractor will figure cost of materials, then charge you for the materials plus a predetermined hourly rate for the actual time he or she spends on the job. Cost-plus bids are commonly used in situations where the work is so complex or tricky that a firm bid would have to be very inflated to cover all the contingencies. If you decide to go with a cost-plus bid on any part of your project, be sure to establish a ceiling above which costs will not rise.

Contracts

A written contract is the best safeguard you have against future misunderstandings, and, should you become involved in a dispute with one of your subcontractors, the contract may be the only piece of evidence available to determine the intent of the parties at the outset of the job. A good contract should be written in the best interests of both parties. Trying to "put one over" on a tradesperson who's working on your home is asking for trouble. The purpose of a construction contract is to prevent the kinds of miscommunications that can result in lawsuits.

For straightforward jobs, it probably isn't necessary to hire a lawyer to write the contract, but it is a good idea to have your attorney take a look at large or complex contracts before you sign them. Many subcontractors have a standard form that they use, or you can come up with your own form. Remember that you can always attach addenda to a subcontractor's contract if you are uncomfortable with it for any reason, or you can delete offending phrases and have both parties initial the change. Both parties should initial and date any contract modifications.

A contract in the interest of both parties should identify:

- The contracting parties, giving names and addresses.
- The location of the job.
- The nature and scope of the work. You can state that the work will be performed according to attached plans and specifications. You might also indicate that the work must be done to the satisfaction of the local building officials.
- The total value of the job and the method of payment. A common arrangement is to pay in thirds: a third when the work starts, another third at a predetermined midpoint, and the final installment when you and the building official are satisfied with the work. *Never* make the final payment until you are fully satisfied. Withholding payment is often the only leverage you have with a sub.
- The starting and expected completion dates. If there are unforeseen and unavoidable delays, the contract can be extended to accommodate them. Often a sub won't accept a penalty clause, so a stipulation that the sub will work continuously, weather permitting, until the job is done probably makes more sense.
- Insurance. It is critical that you require verifiable evidence of insurance coverage from each of your subcontractors. The laws vary from state to state, but in most areas, it is your responsibility to compensate anyone injured while working on your property. In the case of serious injury or death, the dollar amounts can financially ruin the average person. To be on the safe side, many owner-builders take out a blanket workman's compensation policy to cover hourly employees as well as day laborers hired by subcontractors, new employees of subcontractors (they might not be on the subs' insurance lists yet), or anyone else who isn't covered by the subcontractors' insurance. Check with your insurance agent or the workman's compensation agency in your state for more information.
- Responsibility for obtaining permits, inspections, and utility hookups. Some subcontractors, most frequently plumbers and electricians, are used to getting the permits necessary to complete their part of the project. They may also be more comfortable calling for inspections themselves, and are likely to be acquainted with the procedures for getting utilities connected to your home. It may be convenient to have these subcontractors take care of such details for you, but be sure to specify this in your contract, so everyone will be clear about who will be handling them.
- Responsibility for procuring materials and transporting them to the site. Many subs are accustomed to purchasing and arranging for the delivery of their own materials. They

will claim that they can get a better price than you can, since they buy in volume and have working relationships with suppliers. This may be true, but you should confirm it. Ask the sub to break down the bid into labor and materials, then do your own shopping. Even if you decide to have the sub buy the materials, you should pay for them and get a lien waiver from the supplier so the supplier can't put a lien on your property if the sub defaults on any payments to the supplier.

■ Other terms either party wants to include. These might cover the following:

Quality of work. Contracts usually include phrases such as "the work will be completed in a workmanlike manner." Although that may seem vague, it does have meaning in legalese—basically that the work will be done to the standards considered appropriate by that particular trade. If you want to be more specific than that, you should do so in writing.

Responsibility for damage to materials or existing structures. We came across a contract form for contractors that read, in the tiniest print, "The Contractor shall not be responsible for damage to existing walks, curbs, driveways, cesspools, septic tanks, sewer lines, water or gas lines, arches, shrubs, lawn, trees, clotheslines, telephone and electric lines, etc. by the Contractor, subcontractor-contractor, or supplier incurred in the performance of work or in the delivery of materials for the job." If that doesn't seem fair to you, you're right. You should never sign a contract having such a clause.

Responsibility for cleanup. It is remarkable how many disputes arise around the condition of the site both during construction and upon completion of the job. You may be able to save yourself some money and hassle by taking responsibility for cleanup yourself. If you don't have the time or inclination, make sure that the subcontractor realizes that it is a part of his or her job.

Owner recourse in the event of nonperformance or substandard performance. Since it is widely agreed that lawsuits are a no-win situation for all involved, more and more people are specifying arbitration as the preferred method of settling disputes. Name a trusted neutral party to serve as arbitrator.

Your role. If you plan to provide any labor, equipment, materials, or tools yourself, be specific about your intentions.

A stipulation that this contractor accepts the condition of the work that went before. For example, if tiles start coming loose after a tile setter has finished his or her work, the tile setter should take responsibility for this, not try to blame it on poor work by the framers who put down the subfloor on which the tiles were laid.

A stipulation that this contractor cannot assign the agreement (hire another contractor to take his place) without the written consent of the owner.

Working with Subs

Theoretically, your job as owner-contractor should get easier once the contracts with the subs are signed. Assuming you've done your homework, the people you hire will be capable of completing the job without much guidance or interference from you. Still, you should visit the site every day to check on progress, and if something looks off to you, do not hesitate to mention it. Everyone makes mistakes, and a responsible subcontractor will thank you.

If you're pretty sure that a subcontractor is not doing the job the way it should be done, stop work and get a second opinion from an expert. In the eyes of the law, if you have misgivings about the work someone is doing but allow him or her to proceed and then refuse payment, you may be considered "contributorily negligent"—in other words, it's your own fault. So if you have a real concern about a tradesperson's performance, speak up. A couple of examples might be illustrative.

One homeowner hired a tile setter to tile the bathtub enclosure in her home, but as the work progressed, it didn't look right to her. She was embarrassed to confront him, however, so she paid him and he left before she got someone else in to look at the job. Normally tiles are set from the middle of a wall to the edges in such a way that the small tiles at the top and bottom are equal in size, giving the wall a symmetrical look. This tile setter had started at both ends of the wall and worked toward the middle, leaving a disorderly, chopped-up pattern. To make matters worse, the wall faces the door to the bathroom, so it's the first thing you see when you walk into the room. By the time the homeowner realized she was right to be concerned, the tile setter was long gone.

Another owner-builder contracted with a mason to build a large masonry wood stove. The mason had built many standard fireplaces, but never a stove of this type. Still, he chafed at having to follow the instructions of the owner. As construction proceeded, the owner felt certain that the plans were not being followed, but the mason insisted that he knew more about these things than the owner. The stove got built, and when it was finished, it didn't work. Again, the owner confronted the mason, who insisted angrily that it worked fine. The owner then called in a series of experts who confirmed his fears that the stove was basically flawed and would have to be torn down.

ACQUIRING SKILLS

Most owner-builders come to the idea of building their own home with lots of excitement and enthusiasm tempered somewhat by their lack of experience and resulting lack of confidence. It seems, though, that the desire to build for themselves overrides any fears and insecurities, and they find creative ways of compensating for their lack of experience. The most effective way, as we've mentioned, is exhaustive research and planning before the project starts.

The first step in your research (after reading this book!) might be to review other relevant literature. There are a number of periodicals, books, and organizations geared to you, the owner-builder. As you read, some ideas, designs, and building techniques will be particularly attractive to you, so take careful notes and organize them so that the information will be readily available when you want to refer back to it.

There are other sources of information more specific to your locale, such as the local building department, utility companies, builders and subcontractors, materials suppliers, solar energy associations, and energy extension offices. You'll discover that much of the information these organizations offer is free for the asking. Industry trade associations and government agencies are also excellent sources of information for the cost of a letter or phone call.

When you visit model homes, take a 25-foot tape measure and a notebook and keep close track of your impressions of different materials and designs. Often small spaces will appear larger than they are because of the colors or textures used to finish them, the placement of windows, the design of the space (open designs tend to look larger than those with partition walls), or the height of the ceilings.

If their lenders are not pressing them to move fast, some owner-builders take the "learn as you go" approach. They research the project thoroughly and participate in the design process, but come to the construction phase without much experience of such work. They either hire professionals as consultants or work alongside them during the stages they are particularly unsure about. Or they tough it out themselves, making up for their lack of expertise with time and careful attention to the job at hand. Another route is to hire an expert on a consulting basis to help with planning the job and then to check your work once it's finished, leaving you to perform the mechanics of the job.

Owner-builder schools in many parts of the country offer classes, seminars, and workshops in housebuilding and related subjects. A complete list of these organizations is provided in Appendix B. Most owner-builder schools also act as informal referral centers to put owner-builders in touch with each other, government agencies, or other organizations that might be useful to them—receptive lenders, trustworthy real estate agents and tradespeople, and other sources of information and materials.

If there are no owner-builder schools nearby, you can still get hands-on experience by loaning yourself to a builder or owner-builder. One arrangement that works well is to trade work with other owner-builders. You help them now and gain some experience, and they help you later when you're working on your house. How the arrangement is structured is up to the parties involved, but assuming you are both at approximately the same skill level, trading hours one-for-one usually works.

Another way to get a feel for construction work is to plan and execute some projects in your current home, for example a small remodeling job in one room. If you don't own your home, you might ask the landlord to purchase the materials if you do the work. If your landlord isn't receptive or the house just doesn't need any repairs or improvements, ask family, friends, and neighbors if they have projects in the offing.

You are always better off working with someone who has a higher level of expertise than you do, if that's possible. If you take on a project alone, don't commit yourself to a rigid schedule. You'll want to be able to take your time and give careful attention to each phase so that it is a genuine learning experience. Ideal projects are things like finishing basements or spare rooms, where the mess caused during construction can be confined to one area.

POTENTIAL PITFALLS

Perhaps the greatest benefit we can give you is to help you steer clear of problems that other owner-builders have run into. Here's a summary of the most prevalent pitfalls.

Timing and Scheduling

As an owner-contractor, setting up schedules is your responsibility, whether you are hiring out all the work or doing some of it yourself. The single most common and costly error novices make is not allowing enough time for planning. You will be saving the most money at that stage, since you will be anticipating and resolving difficulties before the building starts.

In order to keep things running smoothly after construction begins, you must have a grasp of the logical and orderly progression of a building project. Obtaining this knowledge is a critical part of your planning. See the PERT charts in Appendix A for the order of events in the construction of a typical home. In addition to the PERT charts, the work-breakdown form presented earlier in this chapter will help. It allows you to track what's happened so far and compare the actual time various jobs have taken to how long you had planned for them. Thus, you will know right away if things are falling behind, so you will be able to contact the affected subcontractors and suppliers and reschedule their jobs.

Not being assertive with subs is a mistake that we see owner-builders make over and over. Many subs believe in the "squeaky wheel" concept, and if you're not willing to play the squeaky wheel, you may not get your job done on schedule. So be friendly but firm. One woman who built a small house in the mountains west of Denver scheduled an excavator to dig the foundation and driveway. He assured her that he would get it done that week, and she blithely went off for a week, assuming that the excavating would be done when she visited her site on the weekend. Not only was it not done by that weekend, but she and the excavator repeated that little drama for the next four weeks. He always seemed to have plausible excuses, and she had little choice of excavators in that part of the world. Since winter was coming on, she was getting a little anxious. Finally, she started paying daily visits to his home for almost a week until he finally got around to doing the work.

Strain on Family Relationships

It is useful to realize at the outset that a housebuilding project can add to the day-to-day strains of family life. The best defense is to resolve to be a little more patient and understanding with each other.

One therapist suggests having the family sit down regularly in the planning stages and discuss each individual's ideas for the new house. Keep track of the similarities and differences in family members' expectations. The therapist also suggests having a neutral person present to offer an objective voice. Allowing all family members to express their views in the design process is a wonderful way to bring them all into the project, and this may help resolve conflicts if they arise later. Even children are less apt to disrupt a process if they feel they have an active part in it.

If differences do arise, separating the people from the issues at hand will help you reach a resolution. You know that you love your husband or wife, and you both want to see the house built. Keeping these things in mind can help you reach agreements that you both can live with. A good resource for conflict resolution skills is *Getting to Yes: Negotiating Agreement without Giving In,* by Roger Fisher and William Ury of the Harvard Negotiation Project (New York: Penguin Books, 1981).

Underestimating Costs

This seems to be endemic to owner-builders. We suspect that the native optimism owner-builders possess is at least partly responsible for the tendency to underestimate the real costs involved in a housebuilding project. You should check and double-check all your estimates, and when in doubt—go up. Assume that everything will be at least marginally more expensive than you have any reason to believe. It's far better to build a cushion into your budget than to run out of money before the house is habitable. Most lenders will help you by insisting that you give yourself a healthy margin for error before they make the loan to you. But if you're borrowing from friends or family, or building out of pocket, you'll have to police yourself.

Generally, people are pretty close when they estimate the cost of constructing the house itself. But they often overlook such secondary costs as engineering fees, survey

Table 2-2 LABOR/MATERIAL/EQUIPMENT BREAKDOWN

Job	Percentage of Total Cost		
	Labor	**Materials**	**Equipment**
Excavation	59	11	30
Concrete footings	62	38	8
Concrete-formed walls	57	38	5
Concrete floors	65	28	7
Concrete block	42	54	4
Veneer masonry	43	52	5
Framing carpentry	20	80	0
Foundation insulation	31	69	0
Insulation	41	59	0
Roofing	39	61	0
Gutters	26	74	0
Plumbing	53	47	0
HVAC	26	74	0
Electrical	61	39	0
Finish carpentry	40	60	0
Stair construction	25	75	0
Cabinetry (bought)	12	88	0
Drywall	55	45	0
Exterior painting/Sealing	74	25	1
Interior painting/Sealing	72	28	0
Ceramic tile	27	73	0
Floor covering	26	74	0
Landscape	34	63	3

costs, the price of wells and septic systems, road costs, the expense involved in site clearing, etc. By far the most common mistake is to simply leave something out. So pore over your PERT charts to ensure that you don't make this mistake.

Fortunately, self-policing is well within the abilities of most owner-builders. As profiled in the *Rodale's New Shelter* survey, owner-builders are an enterprising, pragmatic, motivated group of people. Although most successful owner-builders have stories to tell about the problems they had to overcome while building their houses, they also speak of the satisfaction, increased self-esteem, and pride that came with creating their own shelters. The opportunity to live in a customized environment, designed to satisfy the aesthetic, psychological, and practical needs of your own family, is well worth the hours of hard work that go into making the idea a reality.

CHAPTER 3

Designing Your Own Home

As you begin to plan your home, you should assess yourself to see whether you have the skills needed for developing your own design. If you feel the need, you can always hire a professional designer to assist you or to take over the bulk of the design work. The success of your building project will ultimately rest on the design you follow, so you should do whatever is necessary to assure that you have a design that fulfills the desires you have for your future home.

You should spend plenty of time thinking about the design of your home, even if you decide to hire a professional. Thinking through exactly what you want will facilitate communication between you and the designer, saving you time and money. Of course, you may not need the assistance of a design professional. Just as there is a symphony conductor in each of us, so there is an architect! Our overspecialized society has allowed that architect to fall asleep so that we have forgotten how to design. Our task in this chapter is to reawaken the latent architect and put him or her to work.

Even the best architects began at a point in life when most of their design skills were latent. Given time, training, and some practice, you could turn out to be the new "Grandma Moses" of American home design! And you don't have to sign up for a six-year Master of Architecture program to get the proper training. If you check around, you'll find less time-consuming design courses and programs. A partial list of programs to investigate includes the following:

■ Owner-builder schools: Besides learning from the instructors, you will also benefit from meeting novice owner-builders who may have helpful insights into your own questions and concerns.

■ Vocational-technical schools: They offer good courses in architectural drafting, cost estimating, building systems, building codes, and computer design.

■ Adult education programs of various university schools of architecture: Although the design courses at these schools are intended for architecture students, you can request admission from the instructor, who may need more students to fill his or her quota.

■ Other continuing education courses offered by colleges in your area.

ESTABLISHING DESIGN GOALS

Before you get your pencils and drawing paper out to begin sketching ideas for your home, sit down and, with a focused mind, set some goals for the design process. What are you trying to do as an owner-designer? Are you going to learn as much as you can about the design process so you can more effectively define your need for professional design assistance? Do you want to be better equipped to communicate with the design professional? Will you try to design the home without a professional? Do you want your home to be a shining example of an innovative building technique, and as energy efficient as you can possibly make it? Or do you just want to create comfortable, affordable living space for yourself and your family?

Quality of design should always be at the top of any list of design goals. How you achieve high quality depends for the most part on your level of skill in the following steps, which we will discuss in detail:

1. Establishing a sound "design program."
2. Developing a general "design concept."
3. Translating the design concept into construction documents.

STEP 1: ESTABLISHING A DESIGN PROGRAM

A design program (also called a building program) is a detailed written outline of the personal needs that you want your home to meet. The style of the program is really up to you, so feel free to be creative about how you organize this document. It usually has two components: a "program of spaces" and a "program of context."

The program of spaces describes the various spaces in the house and their relationship both to each other and to the exterior environment. Creating this program is a way for you to assess the kinds of spaces you want to have in the house and what functions you need these spaces to perform for you, your family, and guests.

The program of context is an inventory of the physical conditions that must be addressed in order for the finished home to satisfy the wants and needs of the inhabitants. If, for instance, your site is subject to strong winter winds from the west, you would be well advised to minimize the number of windows on the west side of the house, both to reduce the chance of breakage and to reduce heat loss. If there is a lovely view to the west, strategically placed small windows equipped with shutters might be the answer.

While developing these programs, you should gather information about the construction system you plan to use. (Chapter 4 discusses various systems.) As you gain familiarity with a particular building system, you will refine your ability to develop a design appropriate to that system.

The Program of Spaces

In your mind, your future house may be composed of conventionally named spaces such as a living room, kitchen, and bedrooms. Or you may not want to "precondition" the design this way, preferring to give the spaces in the house less conventional designations such as a family information center, food preparation alcove, solarium/spa, parents' retreat, children's creativity studio, and so forth. The names of spaces can also suggest original combinations of functions such as a sleeping/sitting space, great room (sitting, dining, cooking), bedroom/study, and the like.

After naming each space, give a concise description of the specific characteristics you have in mind for the space, its size ("250 square feet" or "10 feet by 25 feet"), its orientation ("located on the south side of house"), particular qualities ("high, well-lit space"), furniture requirements ("contains Aunt Sue's hutch—18 inches by 5 feet by 6 feet"), relationships to other spaces ("is adjacent to sitting area with pass-through to kitchen"), geometrical configuration ("long rectangular room"), and so forth. The more specific your design program becomes, the better your chances for creating a design that recognizes all of your needs and dreams.

The program of spaces should never be viewed as fixed in stone, because in the design process that follows, you may discover that certain conditions in the program are physically impossible to accomplish. For example, there may be requirements for too many rooms facing south, resulting in a house that stretches out like a chain of railroad cars with a large surface area and resultant heat loss. So you may find yourself relocating one or two of the spaces to the north.

The program of spaces will guide you in selecting a specific shape and appearance for your home and for each of the rooms in it. As you see the shapes developing in your mind, you may decide to go back and revise the program of spaces so as to arrive at different shapes. This process may recur several times until you are comfortable with the final shapes.

Following is an example program of spaces for a couple who plan to build a passive solar house on rural land in Colorado. They have not had any children together, but the husband has a teenage daughter by a previous marriage; the daughter visits them on weekends and during the summer.

GENERAL DESCRIPTION: The area of the house will be 2,500 to 3,000 square feet, plus an attached garage for two cars and a tractor. The house will consist of three levels: main entry level, upper level, and basement level.

MAIN ENTRY LEVEL: This level will be, except for the library, essentially one large space with subtle subdivisions; exposed beams and structural elements are desirable.

   Entry: Air-lock entry with coat closet; located on northeast or east side of house
   with access to covered breezeway that connects the house to the garage.
   Great Room: Has the atmosphere of a Spanish adobe room. Contains the sitting
   area and dining area. Has direct access to kitchen (one step up) and the solarium
   (one step down through French doors). Has a wall with built-in bookshelves,

display cases for collections, and stereo/TV systems. Between the kitchen and dining area will be a wet bar island (which, without the stools, will serve as a buffet). The dining table will seat eight. The dining area will overlook an adobe fireplace located in the wall between the table and the solarium.

Library: Has a wall with built-in shelves for books. Two desks. Two reading chairs. Direct access through French doors to great room and solarium.

Kitchen: Close to main entry. Access to dining area through wide arched opening in which is located the island wet bar/buffet. Can be located on the north side of the house. Near to separate walk-in pantry.

Half Bath: Near to main entry and library.

Solarium: A high space joining the main and upper levels. Octagonal, hexagonal, or polygonal floor plan with several sets of French doors to adjoining rooms on both levels. At least half of the square footage of the solarium will project forward from the south wall of the house.

UPPER LEVEL: Three bedrooms adjoin the upper space of the solarium.

Master Bedroom: The bedroom and bathroom/dressing area could be subtly subdivided. King-size water bed (verify weight for structural calculations later). Built-in dresser and bench with drawers in the bedroom. There will be a sitting area for two in the bedroom. The room will be on the east side of the house for morning light. There will be small balconies with wrought iron railings overlooking the solarium and the outside of the house.

Master Bathroom: Walk-in steam shower (two shower heads). Double sink with large mirror over. Tub (6 feet) with a low viewing window (tub could be on raised platform). Water closet in a small space separate from rest of bathroom. Walk-in wardrobe (6 feet by 8 feet) with skylight and full-length mirror.

Bedrooms 2 and 3: Full-size beds. Wall closets. Each bedroom has a separate half bath. Bedroom 2 will be larger than bedroom 3 and will be the daughter's room. Bedroom 2 will have a small balcony with wrought-iron railing overlooking the solarium.

BASEMENT LEVEL: Full basement. All of this level will be unfinished. Rough in plumbing for future three-quarter bath. Cedar closet for seasonal clothes storage. Future bedroom with walk-out access. Workshop and storage areas to be planned.

This particular program of spaces evolved as a three-level scheme. You should be aware that multilevel designs are more complex, because of the necessity to vertically align the upper levels with the lower levels. (For example, rooms having plumbing fixtures should be aligned to keep plumbing costs under control; windows on various levels should be composed to make the exterior appearance of the home pleasing; and the location of other vertical elements such as stairways and clothes chutes must also be considered.) Also the relationship of various levels to each other will determine whether upper-level rooms are cantilevered or whether lower-level rooms project, thus requiring roofs.

(continued on page 38)

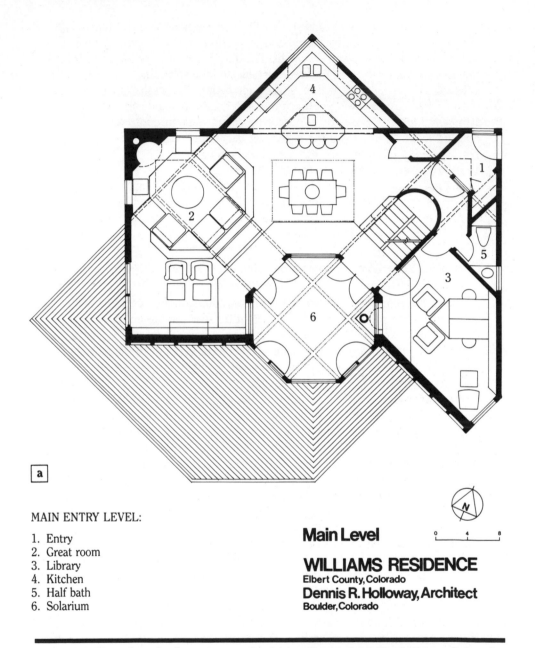

a

MAIN ENTRY LEVEL:

1. Entry
2. Great room
3. Library
4. Kitchen
5. Half bath
6. Solarium

Main Level

0   4   8

WILLIAMS RESIDENCE
Elbert County, Colorado
Dennis R. Holloway, Architect
Boulder, Colorado

Figure 3-1: Floor plans for the main entry level (a) and upper level (b) of a house designed on the basis of the program of spaces presented in the text.

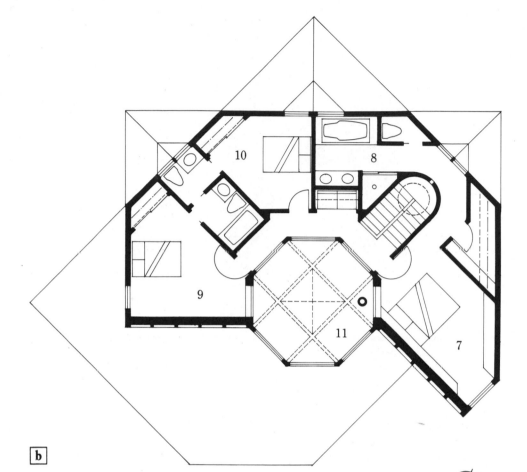

b

UPPER LEVEL:

7. Master bedroom
8. Master bathroom
9. Bedroom 2
10. Bedroom 3
11. Upper solarium

Upper Level

WILLIAMS RESIDENCE
Elbert County, Colorado
Dennis R. Holloway, Architect
Boulder, Colorado

N

0   4   8

Figure 3-2: Exterior perspective of the house designed to fulfill the program of spaces. (Design by Dennis Holloway, architect.)

How you decide whether your home should be one level or multileveled relates to the following considerations:

■ You may have subjective preferences. Perhaps these will be based on memories of favorite houses from your past, or designs you have seen in magazines or books.

■ Energy conservation studies have shown that the most efficient buildings are those with the smallest ratios of surface area to volume. Spreading a house out on one level may result in a large ratio compared to a house that has been organized into several levels. Thus, a two- or three-story home is apt to be more energy efficient than a one-story design.

■ There may be a need to separate certain spaces from each other. Often, this can be accomplished only by locating these spaces on separate floors. For example, a teenager's bedroom may need to be distant from an older couple's bedroom; a workshop may need to be kept far from a reading/study room; and so forth.

The Program of Context

The program of context may be organized in a variety of ways. It is essentially an inventory of all the external conditions that can act upon the house. The following summarizes these conditions.

Major Climatic Factors

■ Solar orientation: The relationship of the summer and winter sun to the house site. For solar heating, one wall of the house should face south.

■ Temperature: The high and low temperature averages for various seasons, night and day. This information will have a direct bearing on how energy efficient you need to make the house.

■ Precipitation: Rain and snowfall at the site (depth of snow is especially important for later structural calculations), and the direction from which snow drifting occurs (you don't want snow blowing in your doors).

■ Wind: The average wind speed and direction for various seasons. In the North, you probably need to protect the house from the prevailing winter wind, whereas you may want to keep the house open to cooling summer breezes. In the South, you may need to open the house to mild winter breezes but close the house in summer when outdoor temperatures soar.

Minor Climatic Factors

■ Trees: The location of shade trees, both deciduous and coniferous; how shadows of trees pass over the proposed house site and consequently affect the potential for solar heating; whether the existing trees can serve as windbreaks to block unwanted winds.

■ Landform: The location of hills or slopes that could block the sun or shield the house from winds.

Physical Site Factors

The importance of spending time on the site to experience its positive and negative features cannot be overstated. The American architect Frank Lloyd Wright was known to camp at a site for several days during different seasons to get to know it intimately before he began the design process. Two procedures that yield information you'll need are the soils test and the topographic survey:

■ Soils test: The purpose of testing the soil is to get precise data to help in designing the home's foundation and footings. If you are certain about the kind of soils you are dealing with on your site, you may be able to forgo the soils test. If, however, there is any question, the money you spend here will be well worth it. This test should be performed by a professional soils engineer. Some land grant colleges in the Midwest have soil testing laboratories and, for little or no charge, they will test samples that you bring in. If you want to use this service, consult a structural engineer about the proper way to take the soils sample.

■ Site topographic survey: The purpose of the topographic survey is to get accurate information about the change of level on the site—where the high and low points are; how the land slopes; the location of trees, boulders, utilities, roads, and so forth. The drawings produced by the surveyor can be studied to determine the best locations for the house and driveways or roads. A model of the site can also be built from the information on this drawing; we will discuss this later in this chapter. You can learn how to do the survey yourself by studying textbooks and renting the appropriate surveying equipment. In general, however, surveying is best left to professionals.

Other Physical Site Factors

■ Important views: Do you want certain views to be visible through the windows of various rooms in the house?

■ Access: How will automobiles enter the site and approach the house?

■ Privacy: Is the site hidden from neighbors' view or does it have potential for privacy with additional features such as hedges or fences?

■ Utilities: Can provisions be made for power, sewer, water, well, gas, and other service hookups?

■ Special features: Various rock formations, streams or ponds, landforms, vegetation, and other features may be worth preserving to enhance the aesthetics and livability of the site as a whole.

STEP 2: DEVELOPING A DESIGN CONCEPT

After developing your design program, you should proceed to a general concept of the shape and layout of the house, a concept that integrates all the requirements in the design program. Usually, this design concept is presented as a drawing. Here are a few principles that will help with the task. You may not be able to achieve all of these principles in your first attempt, but with repeated attempts, your efforts will be rewarded.

■ Think about the concept from both the general to the specific and from the specific to the general. It has been said that when architect Eero Saarinen did the first sketches of a

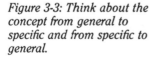

Figure 3-3: Think about the concept from general to specific and from specific to general.

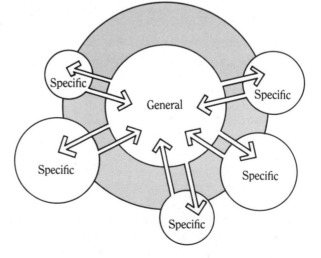

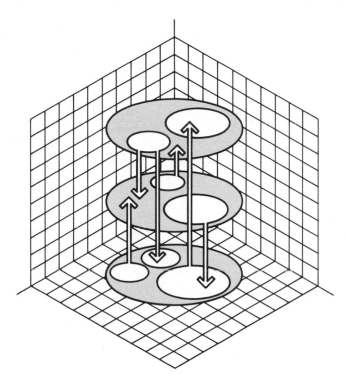

Figure 3-4: Think about the concept from the top to the bottom and from the bottom to the top.

design concept, he showed not only the overall building with thumbnail-size plans and perspective sketches, but also several building details such as windowsills and structural components. The objective was to see how all the parts of the design—the largest and the smallest—fit together to form a whole.

For example, you may want to test a concept that has an overall rectangular plan. Experiment with several different width-to-length ratios of the rectangle to find the shape you like best, then try fitting the various spaces from your program into the different rectangles. Each attempt will teach you something, and experimenting with the arrangement of specific rooms will eventually lead you to make a conclusion about the best overall shape for the house. In moving back and forth from general to specific considerations, the original assumption is modified.

■ Think about the concept from the inside to the outside and from the outside to the inside. For example, you may decide that a certain kind of window would be good for a bedroom and later discover that the building elevation (side view) where that window appears does not look right, and the window creates a visual tension—it lacks proper relationships to other elements on the elevation. Seeing this will lead you to conclude that another window size or proportion would work better. When you are really engaged in the design process, this switching back and forth between inside and outside begins to happen almost automatically.

■ Think about the concept from the top to the bottom and from the bottom to the top. Any juxtaposition of shapes is possible in a multilevel home. Some won't make sense,

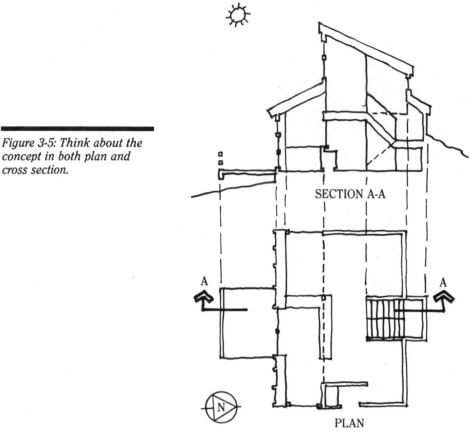

Figure 3-5: Think about the concept in both plan and cross section.

SECTION A-A

A   A

PLAN

though. They may involve costly cantilevers, make the building appear too visually busy, or make the surface area of the home too large. On the other hand, some variety between levels can be aesthetically pleasing. Sameness from level to level may result in a deadly boredom. It is also important to consider how the form of the roof will "reflect" the building plan. It is easy to develop a plan that cannot be roofed in a visually pleasing way.

■ Think about the concept both in terms of floor plans and in terms of cross sections. Architecture novices often limit themselves by viewing their work too narrowly. They see their ideas emerge purely as floor plans, without a feeling for the volume of the spaces created. Or conversely, they see their ideas emerge as cross sections, with a sense of the vertical dimension, but without a real concept of the horizontal dimension. To be a successful designer, one must think of the vertical and horizontal dimensions at the same time.

■ Think about the concept as a system of activity areas (places where people will perform certain activities) connected to each other by circulation spaces (spaces that people will walk through). Thinking about space as though it were a transportation system is one of the hallmarks of design in this century. In the house you are planning, which spaces will be places you travel to in order to perform some activity, and what routes will you take in order to reach those spaces? Look at your house design from an active and passive viewpoint. What happens when you stop, and what happens when you walk? A home with clear routes between various interior spaces is intrinsically more livable.

There should be a strong distinction between spaces to move through and spaces where you stop to perform an activity. For example, if a living room is to be a place for restful sitting or for conversation, there should not be a circulation route through the middle of the room. The living room would be better as a cul de sac—a dead-end space that is walked to, not through. The same could be said of kitchens. If you have ever tried to bake in a kitchen that is set up in such a way that people pass through it constantly, you will appreciate this.

■ Think about your design as if it were poetry. Architecture as an art form gives each of us the possibility of life filled with meaning. As Winston Churchill once put it, "We shape our buildings and then they shape us." The houses we design will most likely exist after we are gone, leaving a record of our dreams for coming generations. As with all art forms, we are free to express our deepest emotions through architecture. Like poetry, there is no limit to what we can say or how we can say it. So the process of design should not bring with it a constricted feeling, but rather a sense of release.

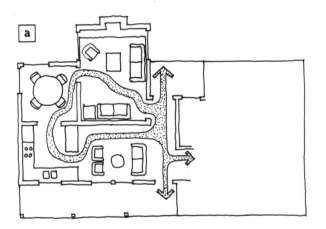

Figure 3-6: Circulation patterns: The location of openings between the spaces can result in circulation patterns that interfere with room activities (a) and patterns that don't interfere (b).

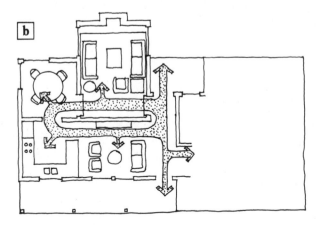

Midway in my life I stopped racing with others. I picked up my dreams and started a gentle walk. My dreams were of a simple house, built with human hands out of simple materials of this world: the elements—Earth, Water, Air, and Fire. To build a house out of the earth, then fire and bake it in place, fuse it like a giant hollow rock. The house becoming a kiln, or the kiln becoming a house. Then to glaze this house with fire to the beauty of a ceramic glazed vessel. I touched my dreams in reality by racing and competing with no one but myself . . . .

Nader Khalili

Racing Alone, A Visionary Architect's Quest for a House Made with Earth and Fire (New York: Harper & Row, 1983)

Design Concepts to Explore

One common mistake owner-designers make is to prematurely finish work on a design concept without exploring alternative possibilities. Such an exploration may lead to a design that will be more satisfying to live in—a design that is intrinsically more open and alive. When suddenly we have the freedom to design our own home, two directions are likely. We either become too conservative, thinking only in terms of houses we have been in, or we go totally radical with fantastic ideas that even Walt Disney would have difficulty realizing. The trick is to strike a balance. The following conceptual explorations may prove helpful to you.

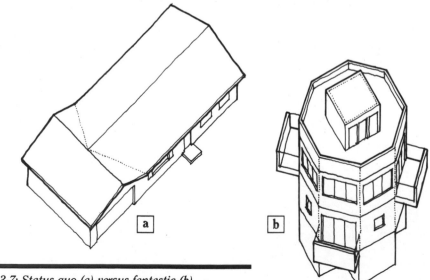

Figure 3-7: Status quo (a) versus fantastic (b).

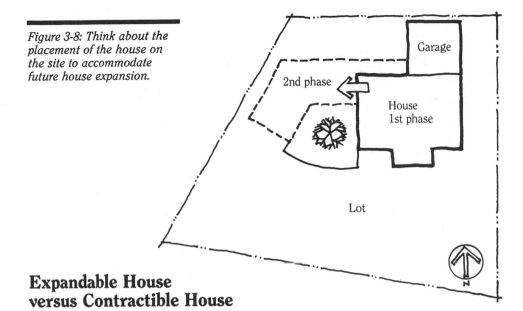

Figure 3-8: Think about the placement of the house on the site to accommodate future house expansion.

Garage

2nd phase

House
1st phase

Lot

N

Expandable House versus Contractible House

Have you ever moved into a house or apartment that was larger than your previous residence? Have you noticed how quickly the new place seems to get filled up with objects so that eventually it too is cramped? It seems we can never get enough space. So we pull up our roots from one place and move on. But you won't want to be forced to move from your new owner-built home just because you didn't foresee how your family's space requirements would change. When the expected baby becomes triplets, or when Mom becomes ill and you decide to care for her in your home, you may wish you had designed more flexibility into the home.

It is useful to think about your design as a central pod to which new rooms can be added. This should be planned carefully, taking into account that a new room cannot be added without affecting the lighting in the existing spaces, or without affecting the circulation patterns of the house. Plan the roof form of the house so that the roofs of additions don't create roof ponding (areas where rain and snow will collect). Situate the house on the lot so that there is space for additions without interfering with lot lines, setbacks, or easements.

You must also consider that the triplets and the visiting parent will not always live with you. When they are gone, what will you do with so many extra sleeping rooms? Can you conceive of a way that sleeping rooms can take on other functions—such as becoming offices, dens, computer rooms, hobby rooms, or informal sitting spaces? Large homes should be able to "contract" in this way (that is, their spaces should be convertible to other uses) when the size of the family decreases.

One-Use Space versus Multiuse Space

There are fundamental differences between spaces designed for one use and spaces designed for multiple uses. Can you feel the difference between your great-grandmother's

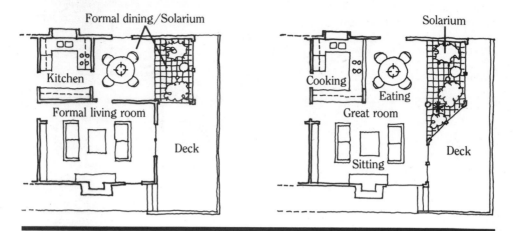

Figure 3-9: Spaces can be designed for one use or many uses.

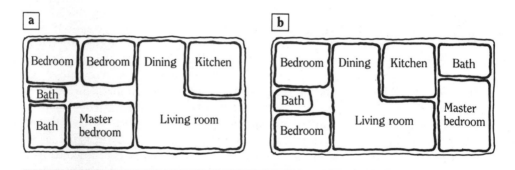

Figure 3-10: Grouping all the bedrooms (a) versus separating the master bedroom from the other bedrooms (b) results in different degrees of privacy.

formal parlor, designed for tea on Sunday afternoons, and a contemporary "great room," designed for sitting, TV watching, eating, cooking, relaxing, reading, and entertaining? Can you imagine a master bedroom that is a space for sleeping, with a writing desk and two reading chairs, and overlooking the solarium below? Contrast that image with a master bedroom that is only for sleeping. Homes with multiuse spaces are more flexible, and often they are more affordable since they can be smaller. A combination living/dining area requires fewer square feet than a living room and dining room.

Space Juxtaposition

Spaces that are next to each other lend character to each other in subtle ways. Imagine a house where a sunspace is adjacent to the master bedroom versus one with the sunspace adjacent to a living/dining room. Imagine an entry space adjacent to the formal dining space, in contrast to an entry space adjacent to the sitting room, in contrast to an entry space adjacent to a stairway hall that leads to living/dining, kitchen, and bedrooms. Imagine a two-story house with the sleeping rooms on the upper story versus a house with the sleeping rooms on the lower floor and the living spaces on the upper story, with good natural light and a commanding view.

Spatial Hierarchy

A house can be composed of spaces of varying importance. The importance of a space can be emphasized by increasing its volume (especially by increasing its height), by furnishing it with attention to detail (especially by using objects that have personal meaning), by emphasizing lighting, by accentuating color, or by any other approach that sets the space apart. Imagine a master bedroom/bathroom with a very high ceiling next to a hallway with a low ceiling—the impact would be great when you enter the room. Make sure the effect you create is intentional and appropriate. Imagine a living room that has a 7-foot-high ceiling next to a dining room with a 12-foot-high ceiling. Does that seem appropriate? Imagine a master bedroom that is smaller than the master bathroom/wardrobe area.

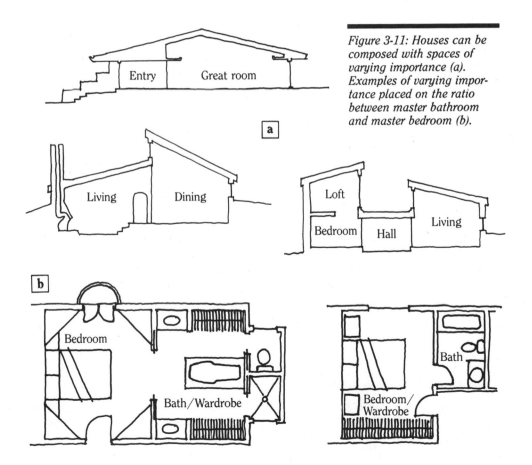

Figure 3-11: Houses can be composed with spaces of varying importance (a). Examples of varying importance placed on the ratio between master bathroom and master bedroom (b).

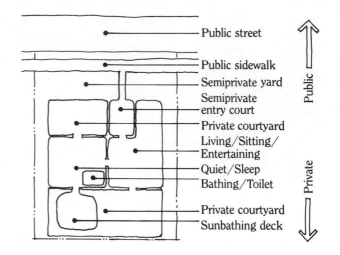

*Figure 3-12: Privacy gradient
for an urban residence.*

Privacy Gradients

Privacy is hard to find in today's world. Your house can be a retreat from overexposure. Be careful that the spaces meant for peace, quiet, and solitude are removed from the spaces that generate activity and noise. Don't place bedrooms near spaces that will be used for entertaining. A similar consideration applies to the exterior of the house. Bedrooms on the street side of the house are less peaceful than those farthest from the street.

As people approach your house from the street, the house should tell them that they are leaving the public realm of the community and entering your private realm. An entrance door in a small courtyard visible from your kitchen or from a secondary living room window communicates more about your privacy than a door that is merely punched into the streetside wall of the house. Stepping up from the street level to the yard and up again to the entry door also subtly communicates privacy.

Spatial Focus

The spaces in most houses were once clustered around fireplaces. But with the advent of modern heating systems, the focus point of the house has become unclear. There is a need to give all houses spatial focus so they don't seem disorganized. The focus can be a family room that is multifunctional or a sitting space surrounded by less important spaces. It can be any area that is clearly marked as the center of life in the home. Don't assume, by the way, that installing a fireplace and hearth will automatically create a focal point. Unless you heat primarily with wood, the fireplace is likely to be just one of many objects in the house.

Open Plan versus Closed Plan

An open-plan house is primarily one large space with subtle space dividers (such as screen walls or low walls) defining subspaces used for specific functions. In contrast, a

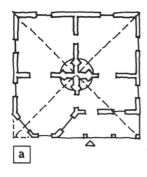

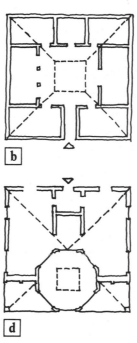

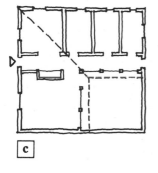

Figure 3-13: Victorian house spaces focus on a central cluster of fireplaces (a). Mediterranean house spaces focus on a central atrium (b). In this contemporary house (c), the spaces arranged as an "L" focus on an exterior-walled courtyard. In this passive solar house (d), the spaces focus on a central solarium.

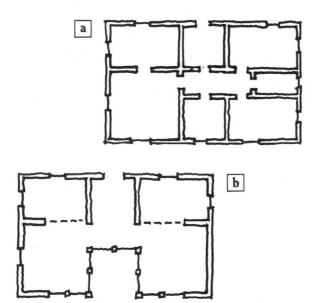

Figure 3-14: Closed plan (a) versus open plan (b).

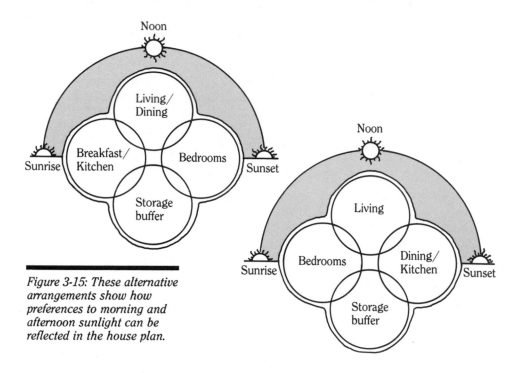

Figure 3-15: These alternative arrangements show how preferences to morning and afternoon sunlight can be reflected in the house plan.

closed-plan house is like a honeycomb—an envelope divided by interior walls into discreet rooms. Both open- and closed-plan designs can be successful, depending on external conditions. For example, a typical English house has a closed plan, because energy prices in Britain are so high that people heat only the room they happen to be occupying at a given time, closing the doors between the heated room and the adjacent rooms. An American passive solar house, on the other hand, usually has an open plan, because the air heated by the sun must be allowed to circulate from the south side of the house through the rest of the house. Although the English example has great privacy, the spaces aren't very dramatic. In the American example, the space is dramatic, but insuring privacy requires careful planning.

Space and Light

So far in this chapter we have looked at design primarily from a conceptual viewpoint—functional relationships, form, circulation, and so forth. There is also a more concrete way to look at design—to look at it literally, through the perception of our eyes. If it were not for light, we would not perceive architectural space or form. While a blind person may have an idea about a space thanks to the senses of touch and hearing, it is with the sense of sight that we can comprehend a space completely. The architect or designer, like the sculptor, manipulates light to define space and form.

The source of natural light, the sun, is a dynamic object, and its motion needs to be understood. The direction from which sunlight will strike the house at various times of day and in various seasons should be taken into account. The color and intensity of light at dawn, morning, noon, afternoon, and dusk should be experienced at the site, and the knowledge applied to the design process.

The house you design has eyes—windows—that allow natural light to penetrate the inner spaces. The impact can be large. You may want lots of light in southern rooms for solar heat, or for reading, for kitchen work, and the like. But some areas—bedrooms or conversation areas, for example—may be more comfortable shielded from strong light. The quality of the spaces in the home will depend to a large degree upon your understanding of light and how you translate that understanding into your design.

Natural and artificial (electric) light can be manipulated by the designer to give very specific qualities to space. The best way to develop a sense of how to do this is to spend time looking at spaces that feel inviting to you and noticing how light is used in them. Look for spaces in which you can study the following considerations:

■ Should there be a relationship between the location of specific spaces and the motion of the sun? For example, should a breakfast room be on the east, south, west, or north side of the house? Do you like to wake up with the sunlight streaming into the bedroom, or do you prefer to go to bed early watching the fading sunset, or do you want a bedroom with light from both directions?

■ What is a space with high windows or clerestory windows like compared to a space with low windows?

■ What is the quality of light from a skylight?

■ Is there a relationship between light quality and the position of window openings in the wall—for example, a wall with a window in the middle compared to a wall with a window at one end? What is the effect of placing a window in one wall adjacent to an intersecting light-colored wall that bounces the light into the space?

■ How does a space that has windows on one wall compare with a space that receives light from windows on opposite or adjacent walls?

■ How do the colors used in the space interact with the light from outside? Do they soften or modify the light? Does the effect change at different times of the day, depending on the position of the sun?

Figure 3-16: Low windows (a) versus high windows (b).

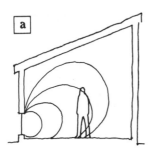
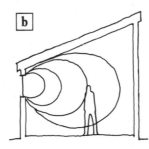

Figure 3-17: Skylights provide distinctly different lighting than do wall-mounted windows.

Figure 3-18: Light from a window in the middle of a wall (a) will support activities in the center of the space while leaving all walls relatively unlit. Light from a window near the intersection of two walls (b) will support activities in the center of the space and create pleasant light reflection from the side wall.

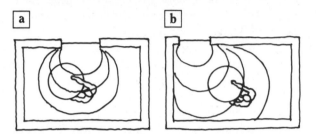

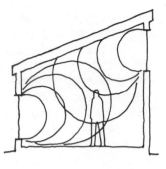

Figure 3-19: Windows on opposite sides of a space produce balanced lighting.

Energy Conservation Concepts

The subject of light leads to the topics of energy efficiency and solar heating. We will discuss these topics in chapter 4, but a brief preview of them is appropriate here. If your design program places an emphasis on energy efficiency or solar heating, you need to consider them when creating your design concept.

In general, unless you will be building in a warm region, the south side of the house should have more window area than any other side, with the smallest window area on the north. But don't forget that some north light is necessary to balance the southern light in rooms, and windows serving as emergency exits may be needed in north rooms to meet local building codes. If a room's windows are so small that electric lighting is necessary during the daytime, you may defeat your energy conservation goal. Since late afternoon sunlight pouring through west windows tends to overheat a room, reduce the size of these windows or shade them with trees, shrubs, or a wide overhanging eave.

To avoid loss of heated air, an air-lock entry or double-door foyer should be used at the entrance to the house. A coat closet can be placed in the entry unless your space requirements for the whole house are stringent, in which case the coat closet can be located elsewhere, keeping the space for the air lock minimal. Closets and storage spaces located on the north side of the house in northern states (or on the south side in the Deep South states) act as "thermal buffers," effectively increasing the insulation value of the wall.

If you are building on a hillside, you can save energy by building insulated walls below grade into the hill—effectively reducing the exposed surface and air infiltration losses of the house. If your site is flat, you can achieve the same conserving effects by "berming" earth around the north, east, and west walls.

Passive Solar Concepts

Elements most commonly used in passive solar homes to make maximum use of the sun's heat include direct-gain windows, direct-gain glazed solariums, and indirect-gain

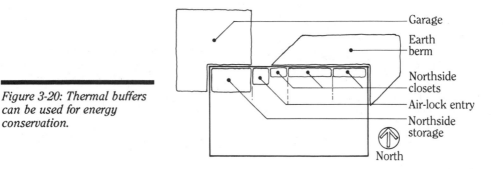

Figure 3-20: Thermal buffers can be used for energy conservation.

— Garage

Earth berm

Northside closets

Air-lock entry

Northside storage

North

Trombe walls and mass walls. Each of these elements will influence your design because they have specific requirements.

■ "Direct-gain" windows allow sunlight to enter the home directly. Much of the heat from the sunlight should be absorbed by some type of high-density material such as masonry; after sunset, the heat will flow out of this "thermal mass," helping to keep the house warm. Direct-gain windows should be oriented due south, although the orientation may be varied by as much as 30 degrees without losing much efficiency. Southerly views from the building site become an important criterion in site selection—you don't want huge southern windows showing you unattractive views. Because many furniture fabrics and carpets are susceptible to fading in sunlight, and because these materials tend to prevent the light from reaching masonry floors where its warmth can be stored, you should keep such fabrics out of direct sunlight.

■ The direct-gain solarium (otherwise known as a solar greenhouse or sunspace) is similar in concept to the direct-gain window, and the same orientation rules of thumb apply. The typical early solarium of the 1970s projected out from the house, like an addition, and was glazed on the south, east, and west sides as well as the roof. The south wall was typically sloped. Today's solarium has been modified for greater efficiency and typically is flush with the south wall of the house, thereby eliminating the loss of energy from the east and west walls. Surrounded by other spaces, the solarium space can be an effective focus for the house, functioning like a solar "hearth." To minimize the overheat-

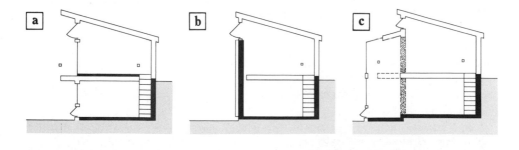

Figure 3-21: A large south-oriented glass wall with low and high vents (a). A Trombe wall (b). A two-story sunspace (c). Thermal mass is shown as solid black and speckled areas.

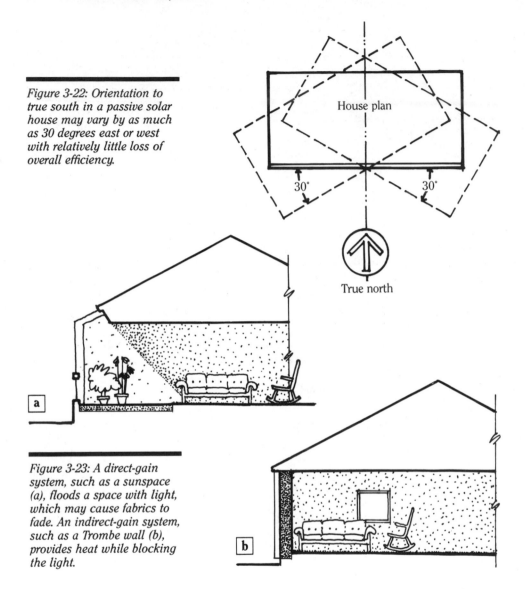

Figure 3-22: Orientation to true south in a passive solar house may vary by as much as 30 degrees east or west with relatively little loss of overall efficiency.

House plan

30°   30°

True north

Figure 3-23: A direct-gain system, such as a sunspace (a), floods a space with light, which may cause fabrics to fade. An indirect-gain system, such as a Trombe wall (b), provides heat while blocking the light.

a

b

ing common in the early style solarium, the roof is not glazed and the south wall is vertical rather than sloped. The state-of-the-art solarium is sometimes a two-story space, with French doors opening to rooms on both levels, allowing better circulation of solar-heated air throughout the house.

■ A Trombe wall is a masonry wall with glazing spaced a few inches outside it. Solar heat is trapped between the masonry and the glass; it enters the house by migrating through the masonry. Whereas the direct-gain window and solarium are virtually transparent, creating strong spatial connections between indoors and outdoors, the Trombe wall obstructs views to the outdoors, so it works well on a site where a southern view is not desirable. If you do want a south view, however, you can place windows in a Trombe wall. Variations on the Trombe wall include half-Trombe walls with direct-gain windows above, and Trombe walls with integral fireplaces. A Trombe wall can also be "bent" or shaped to fit the internal requirements of a floor plan.

The design of a multilevel passive solar house should take into account the fact that there will be some degree of heat stratification, with warmer upper-level spaces and cooler lower-level spaces. Thus the spaces on the upper level might include the living, cooking, and family activity areas where most of the waking hours are spent, and the lower-level spaces could be used for sleeping. Although this "upstairs/downstairs" relationship seems unconventional, it offers a better view from the living space and is ideal for a hillside house with entry on the north side of the house and the north walls of the lower level sheltered by the hill.

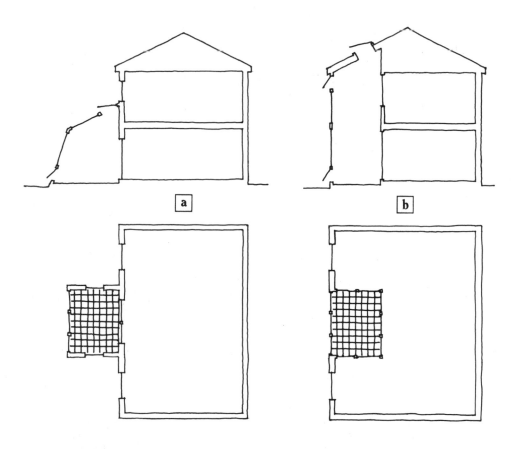

Figure 3-24: First-generation sunspaces (a) usually protruded from the house. New sunspaces (b) are often two-story designs set into a house's south wall.

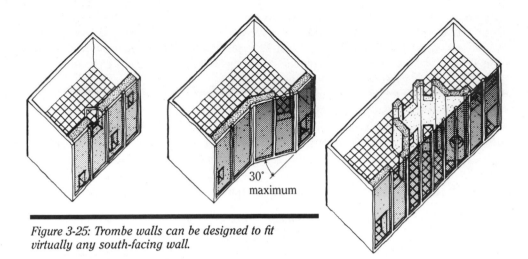

Figure 3-25: Trombe walls can be designed to fit virtually any south-facing wall.

Design Concepts as Real Estate

Owner-designers should think about the design of their houses from a resale point of view. After all, a significant piece of real estate is being created. You may want to sell the house someday, so you should consider how attractive the house might be to potential buyers. There is a fine line between a house that is original, unique, and well designed, and a house that is too idiosyncratic, esoteric, and eccentric. So while on the one hand you want to let your imagination soar with your house design, you may also want to temper your imagination with this sobering truth.

Planning a house with an extra bedroom and bathroom that can be finished before putting the house on the market is often a worthwhile investment. Along the same lines, an unfinished basement with roughed-in plumbing is nearly always a good investment. Designing spaces that can accommodate many different furniture arrangements will also increase the market value of the house—this is especially true of living/dining areas and master bedrooms. Spaces that appear generous and well lit (even if this is a clever illusion) are also popular.

So an important part of the owner-design process is to imagine how others will see your work. In this sense, your design should transcend your personal preferences and become universal.

How to Express a Concept

A "concept" is a thought; it resides in your mind. You must find a way to express this concept in clear, specific terms. This will help you to examine and improve your concept, and it will enable you to express the concept to other people. We will describe several ways to express a design concept.

Drawings

Architect Jacques Bronson said, "Ideas are a dime a dozen; show me the drawings!" Learning to do architectural drawings, even preliminary concept sketches, may require much study, experimentation, and experience on your part. Do not expect to sit down and instantly whip out drawings of your design concept. Realistically, as a beginner, you should expect to spend time to perfect this task—even as much as a year or more. You may have an early "flash of insight" about the form of your design concept, but you will probably work and rework the idea in numerous sketches before something genuinely good emerges.

Design is a patient search; you cannot force it, so relax and take your time. You need to take breaks from creative work like this. Doing something completely different periodically gives your psyche time to recuperate and synthesize the emerging design so that new ideas can flow out of you the next time you sit down to work. Indeed, the best time to get the flash of insight is after you have been away from the design work for awhile. Relaxation exercises may be very beneficial—the act of relaxing while withdrawing the senses, after a period of intense concentration on your design work, will release energy that can be put to use when you resume your work.

There is no set formula for selecting the drawing materials that will best suit your needs and skills. The following is a list of materials that you can experiment with:

■ Pens, pencils, markers: Don't be timid; work in black and bold colors with drawing implements that flow smoothly. Your options include felt marker pens in various nib sizes and colors, nylon-pointed marker pens, or soft graphite pencils (use pencils with 2B, B, or HB leads).

■ Paper: Use inexpensive onion skin or "bumwad" paper (available in rolls of 12-, 18-, or 24-inch widths). Record one idea and then modify it by laying a piece of tracing paper over it, tracing the portions that will be unchanged, then drawing the modified portions. The end result of a stimulating design session will be a nice stack of drawings, in sequence, which you can review later for further refinements and revisions.

■ Sketchbook/Notebook: A three-ring binder with 50 sheets of bond paper is a good place to start. Carry the sketchbook with you wherever you go; design flashes for some of the greatest architecture have been recorded on restaurant napkins because the architect didn't have any other paper handy. You should even place your sketchbook next to your bed at night, in case inspiration strikes during the small hours.

Having selected your drawing materials, you need to decide what kinds of drawings to do. Here are two productive types:

■ Thumbnail sketches: Design ideas occur to us as mental images, then the images travel down one of our arms into the drawing fingers to become a visualized concept. This process of remembering the mental image and then drawing it is very much like waking from a dream and trying to remember it. So it is important to capture these insights with drawing techniques that do not require a lot of time. The thumbnail sketch—a quick, small "rough" drawing—is perhaps the most used preliminary sketching technique, because it is the fastest way to record a design inspiration. As one gains experience, a kind of shorthand develops that is unique to you. You will see much more in your thumbnail sketches than the casual observer. You can do thumbnail plans as well as cross sections, allowing you to deal with the house as a totality.

■ Bubble diagrams: The program of spaces can be analyzed using bubble diagrams. Each space in the home is represented by a bubble, which is free to float arbitrarily around in the unstructured, imagined three-dimensional space. You then start to apply the requirements in your program of spaces to the bubbles, in increasingly more specific ways. As you apply the requirements, you will move the bubbles around, making them larger and smaller, in order to meet the requirements. A general requirement might be "the main floor will consist of the following spaces: great room, kitchen, and entry." A more specific requirement might be "the master bathroom will be to the north of the master bedroom, with a door to the hall to the east." When you have met all the requirements in the program of spaces, the positions of the bubbles should reflect the layout of your proposed home.

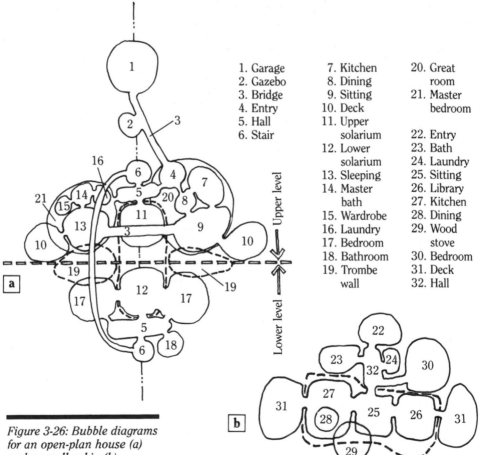

1. Garage
2. Gazebo
3. Bridge
4. Entry
5. Hall
6. Stair
7. Kitchen
8. Dining
9. Sitting
10. Deck
11. Upper solarium
12. Lower solarium
13. Sleeping
14. Master bath
15. Wardrobe
16. Laundry
17. Bedroom
18. Bathroom
19. Trombe wall
20. Great room
21. Master bedroom
22. Entry
23. Bath
24. Laundry
25. Sitting
26. Library
27. Kitchen
28. Dining
29. Wood stove
30. Bedroom
31. Deck
32. Hall

Figure 3-26: Bubble diagrams for an open-plan house (a) and a small cabin (b).

Computers

Personal computers can assist in the design process. Just as computer word processing can increase speed and accuracy in preparing written documents, computer-aided design (CAD) systems help you "draw" your design accurately and with imagery that would take years to master by hand.

CAD systems produce two types of drawings—schematic and scaled—both of which can help you design your home. *Schematic drawings* use symbols or icons to represent objects or processes in the real world. Although signifying the object, the icon may be of different scale, shape, orientation, or color. Bubble diagrams can be effectively rendered using schematic drawings. *Scaled drawings* are employed when an object's true geometrical size, shape, form, and color must be represented. In a scaled drawing, the ratio of one object to another indicates that in the real world, the corresponding objects have the same ratio.

Computers that have great potential for owner-designers include the Apple MacIntosh, Apple II series, IBM PC, TRS-1000, and the Commodore 64. The software you select will depend on the computer you own, since software generally cannot be used on more than one type of computer. MacIntosh is in many ways the premier graphics computer. Software packages available for it include MacDraw, a black-and-white interactive scaled drawing and graphics program; MacPerspective, which permits you to construct wireframe perspective drawings of buildings composed primarily of straight lines; and Mac3D, a 3-D modeling program that hides any lines that would be blocked from view in real life. Many other software packages are also available.

Models

Once you have established your basic concept through two-dimensional sketches or drawings, models can be a useful way to "test" your ideas in the third dimension. This testing process may lead you to modify your design concept. Models can be built to clarify your ideas about interior spaces in the home. And by using solid blocks of material, you can show the exterior "massing" of your concept—how the combined forms of the spaces will appear from the outside of the house.

As your design concept becomes more refined, a structural model (one that shows all the main rafters, joists, and other framing members) can facilitate discussions with a structural engineer. If you have a topographic survey of your site, you can quickly build a 3-D site model by making a separate layer for each grade line shown on the survey. This will help you plan the house in relation to its site.

A house model is usually constructed at a scale of ⅛ inch = 1 foot (⅛ inch in the model equals 1 foot in the finished house), but there is no hard-and-fast rule. Smaller scales will take less time and material and can be constructed in a small work area. A larger scale will permit more detail and, in the case of a structural model, will allow you to test your ideas in depth. Materials for models can be found in art supply stores, model supply stores, and hobby stores. They include:

Figure 3-27: Computer-generated graphics using a MacIntosh computer. (Courtesy of Paul Anderson, architect.)

■ Chip board: A gray/brown paper board in 32-inch by 40-inch sheets, thicknesses of 1/32 inch, 1/16 inch, and 1/8 inch. This material can be cut with a utility knife, is easy to work with, and is useful for both rough and refined models.

■ Illustration board: A gray-core paper board with white surfaces on both sides available in 32-inch by 40-inch or 28-inch by 44-inch sheets, thicknesses of 1/32 inch, 1/16 inch, and 1/8 inch. It can be cut with a utility knife, although not as easily as chip board.

■ Foam-core board: A white beadboard core sandwiched between white paper in sheets up to 48 inches by 96 inches, thicknesses of 1/8 inch, 3/16 inch, and 3/8 inch. It is extremely easy to cut with a utility knife.

■ Wood: Look for basswood before you buy balsa wood, because it is denser and takes a better utility-knife cut. Both woods are available in various sheet sizes in thicknesses from 1/32 inch to 1/4 inch, and there is a large variety of precut pieces useful for simulating structural members or windows.

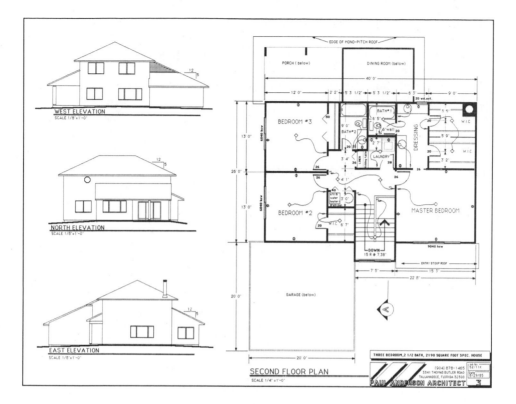

WEST ELEVATION
SCALE 1/8"=1'-0"

NORTH ELEVATION
SCALE 1/8"=1'-0"

EAST ELEVATION
SCALE 1/8"=1'-0"

SECOND FLOOR PLAN
SCALE 1/4"=1'-0"

STEP 3: CREATING CONSTRUCTION DOCUMENTS

After you have established your design program and developed your design concept, you are probably ready to begin preparing the construction documents. If you have never done mechanical or architectural drafting before, you will find formal training useful. The best place to get this training is a vocational-technical school or an owner-builder school. It is not likely that university architecture programs will offer drafting courses, even in adult education programs—such courses are not offered even for architecture majors. Students in those programs are expected to learn drafting while working in architects' offices during their apprenticeship training.

A complete set of final construction drawings will take several sheets (as many as 12 or more), which must be coordinated to thoroughly communicate the design of the whole house. To accomplish this task accurately will require real concentration—definitely not a job to do while watching TV or babysitting! Even a registered architect may take more than two months to prepare a set of drawings.

A piece of advice: Don't begin the drawings and then, for whatever reason, stop work in the project for any length of time. You will find that after a month or two away, you will have forgotten so much about it that you might as well start over. It is for this reason that an architect will insist in his agreement with a client that the project be allowed to proceed without any unreasonable delay. The degree of concentration, mental focus, and memory required for making a set of drawings requires continuous progress in the work.

As the construction drawings progress, you will be taking your concept through its final design development, during which many decisions are made about the various materials and systems to be used in the house. Chapters 4 and 5 deal with these aspects of the process in greater depth.

Required Drafting Equipment

The drafting equipment and materials necessary for making your drawings include the following:

■ Mechanical pencils: Several for holding 2.0-millimeter and 0.5-millimeter graphite leads. The leads should be H, 2H, and 3H (these are measures of hardness), and you should also get nonprint leads (blue leads are good for making notes and lines that won't reproduce when the drawings are printed).

■ Pencil sharpener: A manual sharpener is fine, but an electric sharpener saves time.

■ Drawing board with a linoleum or plastic cover: A good size is 3 feet by 4 feet. If you don't have a desk or table to set it on, you might look for a used drafting table.

■ Straightedge or parallel straightedge (costs more, but preferred): For drawing straight lines.

■ Assorted triangles for drawing angles: Two 12-inch triangles (one for 30-degree and 60-degree angles, one for 45-degree angles), two 6-inch triangles (same angles as the 12-inch triangles), and one 12-inch adjustable triangle (for all other angles).

■ Scales: A 12-inch engineering scale, a measuring ruler, and 6-inch and 12-inch architectural scales. You will use these to measure the construction drawings.

■ Erasers: For various drawing papers and films—ask at the store. Electric erasers are also available and do save a lot of time.

■ Erasing shield (thin metal shield with apertures): You place the shield over your paper and erase through an aperture, so the paper is not smeared while you erase.

■ Circle template, in inch fractions (a plastic sheet with circles of various sizes punched out): Use it to draw small circles and arcs.

■ Drafting compass: For drawing large circles and arcs.

■ Bathroom fixture template (a plastic sheet with cutouts in the shape of bathroom fixtures): Use it to draw these fixtures.

■ Drawing paper or film: Prices vary, but the safest nontear material is "3-mil Mylar matt both sides," available in cut sheets and in rolls.

■ Bumwad tracing paper: In 12-inch and 24-inch rolls for making quick studies for later transfer to final sheets.

Typical Set of Drawings

A set of house drawings usually consists of the following. We have indicated the information that should be included on each sheet.

■ Sheet 1: Plot plan at engineering scale (1 inch = 20 feet), showing the location of house on the property; magnetic north and true north; legal description of the property; contour lines from a topographic survey at 1-foot intervals; planned grading modifications with existing grade contours shown as dotted lines; finish floor levels of the house relative to existing grade; location of utilities and connecting lines; septic field; setbacks and easements.

■ Sheet 2: Footing and foundation plan at ¼ inch = 1 foot scale. This sheet shows the conditions beneath the lowest floor level: all structural wall footings, pads, and pier footings; structural cross sections and details of footing and foundation walls (most of the information on this sheet will be traced from your engineer's "redline drawings"—see sheets 9 and 10, below); footing drain tile; perimeter wall insulation; overall and minor dimensions.

■ Sheets 3, 4, and 5: Floor plans for lower, main, and upper floors at scale ¼ inch = 1 foot. The floor plans are the most important elements in the set of drawings. They are also the most time-consuming to create, containing more information per square inch than the other drawings. Basic decisions about the entire house are made as these drawings are produced. Show all walls, partitions, and openings for doors and windows; all structural supports; kitchen layout with locations for all appliances, cabinets, and counters; bathroom layout with locations of counters and fixtures; overall and minor dimensions; roof over-hang; section and detail cuts. Identify all materials for wall, floor, and ceiling construction; all windows and doors (with door swings); all spaces.

■ Sheets 6 and 7: North and west elevations (Sheet 6) and south and east elevations (Sheet 7) at scale ¼ inch = 1 foot. The elevations are scale drawings of the house's exterior. Identify all building materials, windows, doors, and chimney details. Show exist-ing and new grade lines; below-grade foundation and footing elevations; finish floor lines; metal flashing, gutters, and downspouts.

■ Sheet 8: Building section at scale ½ inch = 1 foot. Other than the floor plans, this drawing will be the most useful one in the set. The section is a view of a slice through the house. Its primary use is to show the vertical dimensions of the house and the way the structural elements are integrated.

■ Sheets 9 and 10: Floor and roof framing plans at scale ¼ inch = 1 foot. The floor plans, section, and elevations are reproduced and discussed with a structural engineer. The engineer studies the soil test report that you provide and makes calculations to determine the size of structural members needed for the house to support its own weight (dead load), the weight of people and other objects in the house (live load), the weight of snow on the roof (snow load), and the pressure of winds (wind load). He will then make what are called "structural redlines" over your set of building prints (these redline drawings will satisfy the local building code requirements). You will use these as the basis of your floor and roof framing plans. You may transfer the redline information onto your drawings or have the redlines reproduced. Show all beams, joists, lintels, columns, supporting elements, connec-tion notes and connection details; overall and minor dimensions. Note special nailing or fastener information; live-load, dead-load, wind-load, and snow-load assumptions. Specify structural grades and conditions.

■ Sheets 11, 12, and 13: Electrical plans for lower, main, and upper floors at a scale of ¼ inch = 1 foot. These plans do not require the services of an electrical engineer. Specify light fixtures (their wattage and location, and the type of switch for each fixture); electrical appliances in the kitchen, utility room, and laundry room; meter location; electrical panel location. Include a "legend" (an explanatory list) defining the electrical symbols used and a "schedule" (an inventory) of the electrical fixtures you want to install.

■ Sheets 14, 15, and 16: Mechanical plans for lower, main, and upper floors at scale ¼ inch = 1 foot. Consultation with a mechanical subcontractor or local power company will give you enough information to complete these drawings. Show location and size of furnace; duct sizes and supply and return-air grille sizes; location and wattage of radiators or other heating/cooling system elements as applicable.

■ Sheet 17: Jamb details at scale 1½ inches = 1 foot. These details are horizontal cross sections showing how the various construction materials are put together at openings, corners, and wall junctures (jambs). Not many notes are required—only enough to indicate materials and some dimensions that are essential to construction. Each detail is given a number, and these numbers are duplicated in circles on the floor plans and elevations to show the locations of the details.

■ Sheet 18: Head and sill details at scale 1½ inch = 1 foot. These details are vertical cross sections through the top (head) and bottom (sill) of the windows and doors showing how materials are put together. The material indications, dimensional information, and reference numbering system are similar to those for jamb details.

■ Sheet 19: Stairway section at scale ½ inch = 1 foot and the cabinet details at scale 1½ inches = 1 foot. The stairway section gives the details for stairway construction. You'll need to study the local building code to assure the design meets code requirements. If you are planning on getting a government-insured loan, check with your lender to be sure that your stairs meet their specifications as well. Cabinet details show how the pieces of the cabinets are put together, and they specify hardware locations and type.

■ Sheet 20: Interior elevations at scale ½ inch = 1 foot. Elevations are usually drawn of each wall of the kitchen, bathrooms, and any other important walls such as shelving walls or fireplace walls. Materials, cabinetry, carpentry, window locations, hardware, and all fixtures are shown.

Needless to say, if you plan to build a one-story house, you will not need all of these drawings, and you may find that your particular design can be shown on far fewer sheets than we have indicated here. The advantage to "isolating" the various systems is that subcontractors can concentrate on their specialty in the project. Separating vertical drawings (sections and head and sill details) from horizontal drawings (plans and jamb details) also increases clarity of communication.

Remember that the main purpose of the drawings is to communicate, and although the conventional method of making construction documents is recommended, you are the boss. If you can understand your own system of drawing and can use it to communicate with your subs, and if the drawings are legible to the building official and acceptable to the lender, then you have accomplished your goal.

From time to time during the production of your drawings, you may need to reproduce them for your records to get preliminary cost estimates from subcontractors or to show them to your lender. Later, when the drawings are complete, you will need to make several sets for the various subcontractors, the lending institution, and the local building department.

Historically, reproductions of drawings were called blueprints because they were created through a costly photographic procedure that produced a negative image of the drawing (white lines on purple background). Today "blueprints" is a misnomer. The most common contemporary reproduction technique creates prints with black, blue, or sepia lines on a white background. This technology is much less costly and easier to read than blueprinting. Ask your printer to put a 20 percent background on the prints, as the chemicals used in the paper may fade over time.

WORKING WITH
A PROFESSIONAL ARCHITECT

If you decide an architect's services are necessary for the success of your project, how can you select the right architect? The best way is to first find houses or buildings that have qualities that are pleasing to you. Talk with the owners to find out who the architect is and whether the owners found his or her work satisfactory. Ask them if you can use their names as a reference when you contact the architect, since a good professional will usually ask how you were referred. You'll want to interview two or three architects whose work you admire to get a comparative sense of how they work, how they structure their fees, how soon you can expect to work with them, and when you can plan to start construction. Some architects will charge an hourly fee for the first consultation while others will charge only for services they perform after an agreement is signed.

In general, architects require a written agreement to perform professional services. It may be a signed letter of agreement, but in any event it's likely to be based upon the American Institute of Architects (AIA) Standard Form of Agreement between Owner and Architect. This form was developed over the past 80 years out of the collective experience of owners and architects. It attempts to protect both parties from undue legal harm by spelling out the responsibilities of each party and the manner of interaction between them.

As an owner-builder, you need to understand that the AIA form does not recognize the possibility that the owner, rather than a professional contractor, would be the builder. The historical reasons for this stem from the idea that the architect loses control during the construction phase of the project—the owner-builder has the final word on what is acceptable, and his or her opinions may differ from the architect's goals. Working with a professional builder gives the architect greater assurance that the design will be followed in a professional manner, producing a home the architect can take pride in. Architects' concern about loss of control over their designs is usually the dominant factor in decisions not to work with owner-builders. If in your interview with an architect this turns out to be the case, don't take it personally—go to the next architect. Any architect is likely to require evidence that you have done your homework about owner-building, can read a set of construction drawings, and have a general understanding of the process of construction management.

It is imperative that you take sufficient time negotiating a mutually acceptable agreement with your architect. Although the AIA does not recognize the owner-builder, most architects will provide professional services to you as long as a written agreement can be established that clearly defines responsibilities, methods of communication, and terms of compensation, as well as allowing the architect a degree of artistic control over the quality of the construction work. Owner-builders working with architects are advised to consult an attorney before signing any agreements.

The Architect's Scope of Service

You can hire an architect on either a comprehensive or limited basis. We'll be discussing limited services in more detail below. Comprehensive services, as defined in the standard AIA agreement form, include five phases, from initial discussions to completion of construction. For an owner-builder, these phases would probably be modified somewhat. Here's a breakdown of the five phases:

■ Programming analysis: Based on initial interviews in which you discuss your needs and budget, the architect will develop the program of spaces and the program of context. If you have developed these programs yourself, then the architect's service in this phase is reduced to a minimum.

■ Schematic design: This is usually the first major phase of service, in which the architect creates a design concept based on the programming analysis, visits to the site, and preliminary concept sketches. The design concept is usually expressed as a set of documents including plans and elevations (at scale $\frac{1}{8}$ inch = 1 foot), an exterior perspective sketch, and an optional interior perspective sketch or study model (at scale $\frac{1}{8}$ inch = 1 foot).

A statement of probable construction cost is also developed during this phase. Presentation of the architect's work at the conclusion of this phase is the first indicator of his or her understanding of your needs, so careful study of the proposed design is required at this point.

■ Design development: With your written authorization to proceed, the architect develops a set of drawings that become the basis for the complete set of construction documents (see below) and more accurate cost estimates. Consulting engineers, under the architect's direction, begin to determine the structural, mechanical, and electrical systems of the building, while you, working with the architect, determine the construction materials to be used. Drawings include a site plan (at scale 1 inch = 20 feet), a building section (at scale $\frac{1}{2}$ inch = 1 foot), and floor plans, elevations, preliminary framing plan, electrical plan, and mechanical plan (all at scale $\frac{1}{4}$ inch = 1 foot).

The documents produced during this phase can be used to begin the process of obtaining financing and building permits. In the hands of a shrewd owner-builder, these documents can even be used for the construction of the building, assuming you can work out all the fussy details that the architect would ordinarily resolve in the next phase.

■ Construction documents: Again with your authorization, the architect completes the set of construction drawings and written specifications from which you will construct the house, and another refined statement of probable construction costs is developed.

■ Construction observation/administration: With commencement of construction, the architect makes periodic visits to the site to inspect the work in progress. The architect can be called upon to resolve or clarify unforeseen details, at least some of which crop up in every project. Because the architect will be evaluating the work you do, it is important for you to keep your mind open and your ego under control. Remember, it is all for your own good! Your relationship with the architect during this phase is like that between the orchestra conductor and the composer and can be a very positive experience for both of you. You can also ask your architect to offer an opinion on the subcontractors' workmanship and code compliance and the appropriateness of their billings.

Fees—The Bottom Line

There is no established fee schedule for architectural service; in fact, such a schedule would be illegal in most states. However, for comprehensive service, it would not be unusual for an architect to ask from 8 percent to 15 percent of a proposed building budget. A 10 percent stipulated sum fee is fairly common. For example, if you plan to build a house that when complete would have a real estate value of $100,000, the fee would be $10,000. Remember that if the architect wants your work badly enough, the fee is probably negotiable. The fee is usually broken down by phase in the agreement, which helps you plan your cash flow. Here is a possible breakdown:

■ Programming analysis phase = 5 percent
■ Schematic design phase = 20 percent
■ Design development phase = 25 percent
■ Construction documents phase = 40 percent
■ Construction observation/administration phase = 10 percent

Usually an initial payment or retainer is paid at the time the agreement is signed. The amount of this fee varies, but could range from 5 percent to 20 percent of the stipulated sum. The initial payment is credited to your account and applied to the last fee statements of the project.

There are other methods of compensation, including hourly rate with a limit, and you should discuss these during the first conference.

Limited Service

You may have decided that designing your own house and doing the drawings is just the kind of challenge you need in your life. In that case, you may want to hire an architect to provide only limited services. Here are a few possible scenarios:

■ You might ask the architect to help you arrive at the design program—or to help you with some preliminary conceptual explorations—before you complete your own design and construction drawings. You might also buy some consulting time to get critiques of your work at various phases. In this case, the best way to pay the architect would be by the hour.

■ You might bring in a carefully worked out design program and ask the architect to prepare a conceptual sketch from which you could do your own construction drawings. Compensation would be by the hour.

■ You might perform the program analysis and bring in preliminary sketches or drawings of the proposed house. The architect would proceed with the design development. You would take the construction drawings from there. Compensation could be either by the hour or as a stipulated sum.

■ You might perform the program analysis and schematic design. The architect would proceed with design development through the preparation of construction documents. Compensation would probably be a stipulated sum.

■ You might bring in a fairly developed design (this could also be a published house plan that you have modified), and the architect would do the construction documents. Compensation would be by the hour or stipulated sum.

■ You might bring in design development drawings and ask the architect to develop the construction documents to the point where you can obtain your building permit. Compensation would be by the hour.

Other Sources of Help

For the owner-builder who has searched for an architect and has failed to find one, there are other avenues of assistance:

■ Architectural designers practice architectural design but are not licensed architects. They may have architectural degrees but have not yet taken the licensing exam or have taken it and not passed it. This is not necessarily a mark against the architectural designer—it wasn't until this century that licensing exams were required of individuals who wanted the title "architect." Still, it is important to check into the designer's experience, examining completed homes and talking with past clients to verify that the homes fill the clients' needs and that the designer met their expectations. The fees for an architectural designer will generally be well below those of a registered architect.

■ Architectural drafting services are small groups of draftspersons (some of whom may be registered architects) who specialize in drawing construction documents. It's possible to have an architect or architectural designer prepare your design drawings and then have a drafting service draw the construction documents. However, you should weigh the fees involved in these kinds of combinations with the extra hassle of coordinating the various personalities. In general, the fee for drafting service is not greatly different from the fee for architectural service.

■ Architectural graduate students offer another way to get design and drawing services. You could call the dean of the architectural program at a university near you and ask for the name of the best student in the program. On the plus side, you can expect an enthusiastic effort, since most students know how difficult it is to get a break for independent work. You can also expect a fresh approach to design, since the student is coming out of a state-of-the-art design curriculum. On the minus side, the student may not have had much "real world" construction experience and may therefore lack a sense of practicality. Still, a first-rate student is probably your best bet for saving money on fees and getting reasonably good service.

CHAPTER 4

Building Techniques for Owner-Builders

As you develop your design, take the time to study, evaluate, and compare various materials and techniques of construction so that you will have some sense of the potentials of the various approaches. Each type of construction has its own "vocabulary" of building elements, so your design may be expressed in different ways. In this chapter, we will introduce some "alternative" materials and building techniques, as well as modifications of more conventional ones.

One of the best ways of getting information about these building techniques is by visiting examples of them and talking with the owners. You should supplement these visits with readings from books and construction periodicals, keeping a record of your observations in a notebook.

Budget limitations, site considerations, and aesthetics are all important in selecting a construction system. Pay attention to your intuition as well. You may get a clear insight into the possibility of an earth-sheltered solar house when exploring a steep south-facing slope of your land. If your region is forest-rich, you may see timber framing or log building as attractive approaches. Also think about your choices in the context of aesthetic compatibility with existing nearby buildings. Spending quiet time on your site often brings up the kind of subjective clues that, along with your objective cost and energy analysis, will help you make a final choice you'll be happy with over the long term.

ENERGY-EFFICIENT HOMES

There is some confusion currently about whether or not we are still having an energy crisis. We hear how at present there is an energy glut, that prices are holding or may even decrease, and that new sources of energy will be developed—probably. All of this talk serves to blur some of the issues.

The price of fossil fuels has risen dramatically since the 1950s. Can you remember the "gasoline wars" of the '50s when a gallon of gasoline cost 15 cents? Since then, the price has increased almost 1,000 percent. Our population has, since World War II, increased about 50 percent, while our consumption of electrical energy (generated primarily from fossil fuels) has increased by 600 percent and our total energy consumption has increased

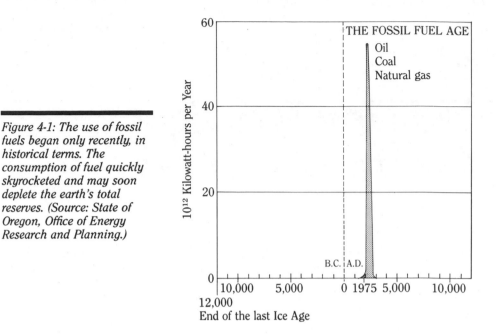

Figure 4-1: The use of fossil fuels began only recently, in historical terms. The consumption of fuel quickly skyrocketed and may soon deplete the earth's total reserves. (Source: State of Oregon, Office of Energy Research and Planning.)

250 percent. In U.S. residential and commercial sectors, each person is using 2 tons of oil annually.

The buildings we live and work in consume about one-third of our national energy budget. It may be a surprise to realize that our present buildings alone consume about twice the electricity that was used for all purposes in this country 30 years ago. New buildings that do not use energy wisely not only increase the pressure on our energy supply lines but also result in more pollution to the planet as a whole. The time is fast approaching when the combustion of fossil fuels will meet strong political resistance as the reality of the "greenhouse effect" in the atmosphere is documented.

The real energy crisis is still at hand, unperceived by most of us because of the masking effect of current economic manipulations. The earth's finite supply of fossil fuels is steadily diminishing. Fortunately, individuals can make a real difference by making informed decisions about energy use in their own homes. Properly planned energy conservation strategies actually pay for themselves in lower utility bills. Most of the work involved in executing these strategies requires little technical skill and is highly satisfying to do yourself.

In our view, all new buildings must be designed with energy conservation as a primary goal. The challenge for owner-builders is to create energy-efficient houses that will make the most cost-effective use of the building systems they choose. A myriad of new building products are now available that can help reduce fuel consumption—we will examine some of these later in this chapter and in chapter 5. The main point to stress now is that there are many simple, commonsense ways to build a home that fulfills all your needs without consuming great quantities of nonrenewable energy.

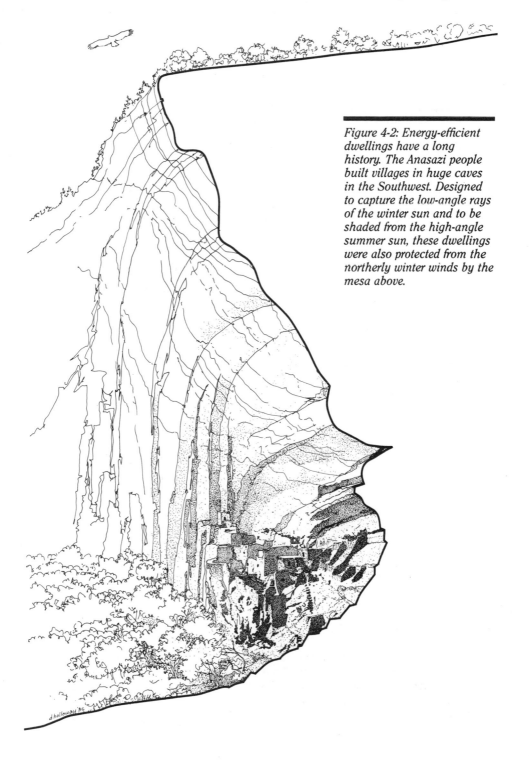

Figure 4-2: Energy-efficient dwellings have a long history. The Anasazi people built villages in huge caves in the Southwest. Designed to capture the low-angle rays of the winter sun and to be shaded from the high-angle summer sun, these dwellings were also protected from the northerly winter winds by the mesa above.

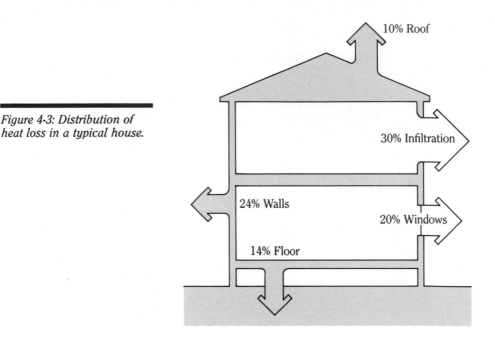

Figure 4-3: Distribution of heat loss in a typical house.

10% Roof

30% Infiltration

24% Walls

20% Windows

14% Floor

THE SUPERINSULATED HOUSE

When we make a house increasingly energy efficient with conventional materials, we reach a point at which the house can be called "superinsulated"—it is far more energy efficient than an ordinary house. Much of what we think of as superinsulating techniques are simply commonsense building methods carried out with an unusual degree of care and attention to detail. We suggest that you take a long look at both superinsulating and passive solar building systems, and use whatever techniques from both approaches that seem appropriate to your situation.

There is a rivalry of sorts between the proponents of superinsulation and those of passive solar construction, with each side trying to one-up the performance of the other. It is fair to say that each probably has something to learn from the other. While passive solar designers have tried to tune their designs to the variations of climate, superinsulation designers have developed a single universal solution for all climates. Both systems are constantly being refined, and the dialogue between them will ultimately help to perfect each.

All superinsulated homes have certain characteristics in common; they can generally be described as follows:

1. Very high levels of insulation are used throughout the building, often requiring thickened walls and other unusual building details.

2. There is less south-facing glass than for passive solar houses. Direct-gain passive solar is considered not crucial to overall building performance.

3. The combined window area for north, east, and west is 50 percent that for south.

Michael Scott, an energy economist in Minnesota, recommends that the total glass area equal 12 percent that of the total floor area.

4. There is no more thermal mass than in conventional construction.

5. The building is virtually airtight. This may require a mechanical ventilation system to prevent the buildup of high levels of indoor pollutants and humidity.

6. The small air change per hour (ACH) achieved in these houses is made possible by a continuous air/vapor barrier. If a plastic sheet is used for this barrier, it must be carefully protected during and after construction.

7. The superinsulated house requires careful workmanship.

8. If all of the above conditions are met, a small direct-vented heater may be the only heating system required.

9. Although these houses are initially more expensive than a conventional house, the extra costs are recovered quickly because of the ongoing fuel savings.

10. The house stays cool in summer if windows are opened at night.

The first thing most people think of when they think of lowering energy requirements for a building is increasing insulation levels. We've found it more useful to think in terms of an integration of a number of elements, all of which must be present and functional in order for the building to perform properly. These features include adequate levels of insulation, an airtight building shell, a controlled ventilation system, a properly sized and designed heating system, and correctly oriented, good-quality windows.

Insulating materials have a high degree of resistance to heat transmission. The relative resistance of a material is expressed numerically as its "R-value," with higher numbers indicating greater resistance to heat flow. Common insulations such as fiberglass, cellulose, and plastic foam insulations usually have R-values ranging from about 3.0 to around 7.0 per inch.

Superinsulated homes have wall R-values as high as 40 and roofs insulated to as much as R-65. To give you some perspective, in many areas local codes only require R-11 walls and R-19 roofs. The insulation levels you choose will depend upon what's available in your area, the degree of energy efficiency you're trying to achieve, the local availability and cost of heating fuels, the economic payback, and personal considerations. We will discuss the types of insulation currently available in chapter 5.

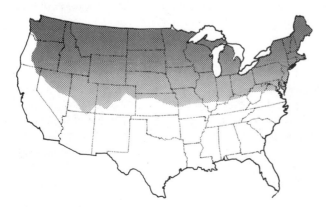

Figure 4-4: This map shows the northern area of the United States where superinsulation may be most beneficial.

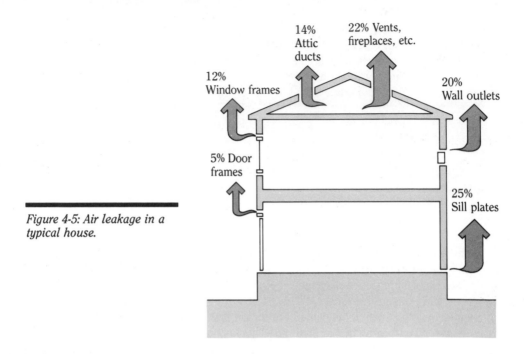

14%
Attic
ducts

22% Vents,
fireplaces, etc.

12%
Window frames

20%
Wall outlets

5% Door
frames

25%
Sill plates

Figure 4-5: Air leakage in a typical house.

Reducing Air Leakage

In most conventional houses, air leakage is the single largest component of heat loss. Factors that influence air leakage include wind speed, the surface area of the house, the temperature difference between the inside and the outside of the house, and the area of cracks and other openings in the exterior surfaces. In conventional homes, it is not uncommon for the entire volume of air in the home to be replaced by outside air .5 to 3 times per hour (.5 to 3 ACH).

Strict quality control is essential to effectively reduce air leakage. One of the most common strategies is to carefully install an air/vapor barrier on the warm side of the wall (under the finish surface) to prevent indoor air and moisture from escaping. Historically, at least 6-mil (a mil is $\frac{1}{1000}$ of an inch) polyethylene sheeting was used for this purpose, but any material that resists moisture flow will work. The relative resistance to moisture flow is referred to as "permeability" or "permeance" and is expressed in perms. The higher the perm number, the more permeable the material. Any material with a permeability of 1 perm or less is considered an adequate air/vapor barrier.

Whether you use polyethylene sheeting or one of the new cross-laminated air/vapor barriers, proper installation involves sealing every joint with a flexible sealant, usually acoustical sealant. Where splices must occur, overlap the sheeting a distance at least equal to the space between two studs, rafters, or joists. Seal carefully with acoustical sealant and staple the sheeting to the studs, rafters, or joists. Then the finish surface (such as drywall) can be installed.

Some kinds of insulation, notably sprayed-in-place polyurethane foam and cellulose sprayed in place with a binder, form their own air/vapor barrier, if properly installed.

Generally, the air/vapor sheeting can be eliminated if you plan to use one of these insulation materials.

In addition to the air/vapor barrier installed on the warm side of the wall, many builders install an air barrier on the outside of the wall, over the sheathing and under the siding, as extra protection against air leakage. This material should have high permeability, so that it will readily allow any moisture that might get trapped in the wall to escape to the outside. Products such as Dupont's Tyvek and Parsec's Airtight White are used for this purpose. The choice of wall sheathing also can reduce leakage. Butt joints between sheathing panels can let outdoor air leak into the house. Insulative sheathing having overlapping joints inhibits this leakage yet permits the escape of indoor vapor that gets past the inside vapor barrier.

In frame houses, there are certain points in the shell that are difficult to seal. Fiberglass or foam sill sealer placed between the sill plate and the foundation wall is an inexpensive way to prevent air leakage at this joint. Caulk or foam gaskets used at joints in the framing will also increase the airtightness of the structure.

Polyurethane foam caulking should be used at all penetrations, including electrical and plumbing penetrations. Take special care when applying foam around door and

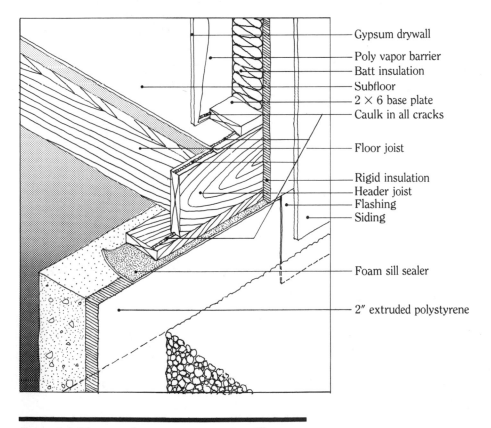

Gypsum drywall
Poly vapor barrier
Batt insulation
Subfloor
2 × 6 base plate
Caulk in all cracks

Floor joist

Rigid insulation
Header joist
Flashing
Siding

Foam sill sealer

2″ extruded polystyrene

Figure 4-6: Close-up of an energy-conserving sill.

window framing—if too much foam is sprayed into the opening between the framing and the jamb, the expansion of the foam can distort the jamb, making the door or window difficult or impossible to operate.

Doors and windows should be weatherstripped. With the relatively small cost of an additional door, a double-door "air-lock" entry can be built to reduce leakage. Sliding doors are leaky and should be avoided; several companies now manufacture swinging "atrium" doors that have the appearance of the old sliders, but are much more energy conserving. Windows in general should be the swing-out type (casement or awning), because their locking mechanism seals the weatherstripping between the sash and the frame. Sliding type windows (sash and sliders) should be avoided unless their measured leakage rates equal those of casement and awning windows.

Other spots prone to air leakage include the places where interior partitions meet outside walls and ceilings. At these spots, pieces of vapor barrier should be secured before the partition is erected, and later the pieces should be lapped into the full wall vapor barrier. Exterior corners can be caulked (sheathing to framing) or taped.

Joist headers present a particular problem in terms of keeping the air/vapor barrier continuous. Some builders terminate the air/vapor barrier at the joists and install pieces of rigid foam insulation with a low permeability in each cavity between the joists. Another approach is to run the air/vapor barrier around the joist header and over the face of the floor below. It is critical that the joist header be set in from the edge of the wall and insulated on the outside, since the air/vapor barrier must always be on the warm side of the insulation.

Superinsulated Stud Walls

The following are some specific wall details that have been used in superinsulated homes. Whether or not you decide to superinsulate, the techniques shown here should be useful to help you think in terms of making your home as energy efficient as is feasible.

■ 2 × 6 wall with foam sheathing: 2 × 6 construction is used with R-19 fiberglass batts and a foam insulative sheathing. The foam could be urethane, isocyanurate, expanded polystyrene (EPS or beadboard), or extruded polystyrene. The insulating sheathing covers the entire exterior of the wall, including the rim joist.

Variations on this wall include the "Styro House," which features 2 × 4 frame construction with R-11 fiberglass insulation and 5.25 inches of EPS on the exterior. The air/vapor barrier is installed on the outside of the sheathing on the frame wall and the EPS is applied over that. It is safe to install an air/vapor barrier inside a wall as long as two-thirds of the R-value is on the outside so that the air/vapor barrier stays warm. This wall design was developed by Bill Brodhead of Buffalo Homes, Riegelsville, Pennsylvania.

■ The strapped wall: This is similar to the 2 × 6 wall described above, except that the wall is furred out with 2 × 3s applied horizontally to the interior of the studs. The air/vapor barrier is installed over the studs and under the 2 × 2 strapping. The space created by the strapping can be used as a wire chase, eliminating the need to penetrate the air/vapor barrier for electrical boxes. The space can also be filled with insulation.

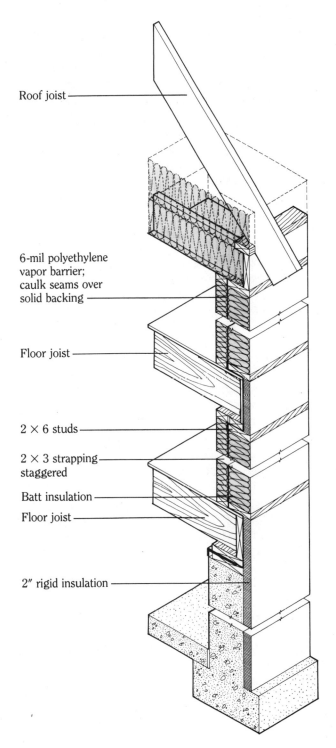

Roof joist

6-mil polyethylene
vapor barrier;
caulk seams over
solid backing

Floor joist

2 × 6 studs

2 × 3 strapping
staggered

Batt insulation

Floor joist

2" rigid insulation

*Figure 4-7: Strapped wall
superinsulation construction.*

■ The double wall: This wall is actually two walls with a space between them. The air/vapor barrier is placed on the exterior of the inside wall, and the interior wall, the exterior wall, and the space between them are all filled with insulation. This allows an R-value of 45 and creates a 3.5-inch cavity for plumbers and electricians to work in without violating the air/vapor barrier. The inner wall is the bearing wall (it carries the structural load), so the exterior wall needn't be designed to carry any overhead weight and it can be built without headers.

■ Staggered stud wall: A wall can be made extra wide through the use of studs staggered on wide plates: Some studs stand along the inner edge of the plate to support the interior finish surface, while other studs stand along the exterior edge for the exterior surface. Gene Leger from southern New Hampshire uses staggered 2 × 3s on 2 × 6 plates, and he fills the walls with sprayed foam. In this design, no vapor barrier is necessary since the foam acts as one. Sprayed cellulose also works well in this situation.

■ Airtight drywall approach: This system uses the drywall interior finish of the house, applied carefully with foam gaskets located in the framing and between the drywall and the framing to create airtight seals. Most moisture that gets into walls is carried there through cracks and other openings. By carefully eliminating all such openings in the drywall surface and applying a good two-coat paint job, you should be able to keep both moisture and air from leaking into the wall. If you want the extra protection, choose a paint marketed as a vapor barrier.

■ Foam-core panels: Foam-core panels were originally developed for use in the refrigeration industry. Consisting of a core of foam insulation (generally polystyrene or urethane) sandwiched between two layers of sheathing, they can be used in timber frame construction: The panels are attached to the outside surfaces of the posts and beams with construction adhesive and large nails. Urethane has a higher R-value per inch than polystyrene, and it acts as a vapor barrier, with a perm rating of about 1 for a 3½-inch core. It is also, however, much more expensive. Since polystyrene doesn't act as a vapor barrier, you may want to paint the interior surface with a vapor barrier paint.

Wiring these walls can be handled a number of ways. Some panels come prewired; others have a chase or thin-wall conduit in the foam under the drywall that wires are fished through. We've also seen some attractive baseboards designed to accommodate wiring, and Wiremold (The Wiremold Place, Consumer Products Division, West Hartford, CT 06110) makes a surface-mounted wiring product for applications like these.

THE PASSIVE SOLAR HOUSE

Using the sun as a heat source is nothing new. Socrates observed 2,400 years ago: "Now in houses with a south aspect, the sun's rays penetrate into the porticos in winter, but in the summer, the path of the sun is right over our heads and above the roof, so that there is shade. If then this is the best arrangement, we should build the south side loftier to get the winter sun and the north side lower to keep out the winter winds."

While the house that Socrates described probably lost heat as fast as it was collected, the Romans discovered that if the south-facing windows were covered with glass, the solar

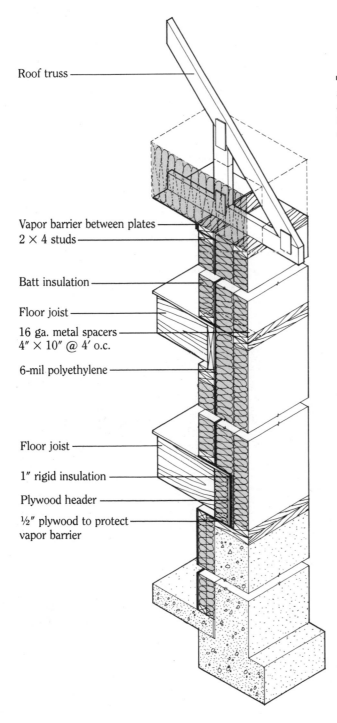

Roof truss

Vapor barrier between plates
2 × 4 studs

Batt insulation

Floor joist

16 ga. metal spacers
4″ × 10″ @ 4′ o.c.

6-mil polyethylene

Floor joist

1″ rigid insulation

Plywood header

½″ plywood to protect
vapor barrier

*Figure 4-8: Double-wall
superinsulation construction.
(Adapted from Roki Assoc.,
Standish, Maine.)*

energy would be trapped, causing the internal temperature to stay high well into the night. This simple phenomenon is called the "greenhouse effect." Today we call a house that uses the greenhouse effect a "passive solar house."

It is a common rule of thumb that, compared to a conventionally designed house of the same square footage, a well-designed passive solar house can reduce energy bills by 75 percent with an added construction cost of only 5 to 10 percent. In many parts of the United States, passive solar houses do not require any auxiliary energy for heating and cooling. Given current and future projected fuel costs, the additional construction cost is recovered quickly.

Like the superinsulated house, this style of building has some distinctive design features.

1. Most of its windows face south. Sunlight passes through these windows and is absorbed by materials inside the house. The heated surfaces of these materials reradiate the heat energy to warm the house.

2. Ideally, the materials that the light strikes are high-density "thermal mass," such as concrete, brick, stone, or adobe. These materials, because of their ability to absorb energy rapidly and reradiate it slowly, can prevent the house from overheating while the sun is up and they help keep the house warm when the sun is down.

3. Architects and builders used to keep window areas on the east, west, and north sides of the house as small as possible. This is still the general rule of thumb, but the

introduction of high-tech window glazings (see chapter 5) lets us relax this rule. Still, house plans with long south walls are best for passive solar houses.

4. Passive solar homes tend to be well insulated and have reduced air leakage rates, keeping the solar heat within the building envelope.

5. Since auxiliary heat requirements are greatly reduced in a passive solar home, small direct-vented units or wood stoves are often the heaters of choice.

6. Passive solar homes often have open floor plans to facilitate the movement of solar heat from the south side of the house through the rest of the house. Sometimes small fans are used to aid warm-air distribution, especially in houses having floor plans that are not fully open.

Passive Solar Techniques

The heating techniques used in various types of passive solar houses are essentially easy to comprehend. In general, all passive solar houses can be divided into two categories: those that use "direct-gain" heating and those that use "indirect-gain" heating.

Direct Gain

Direct-gain houses, considered to be the simplest type, rely on sunlight shining directly into the living spaces through south-facing windows. While some of the heat is

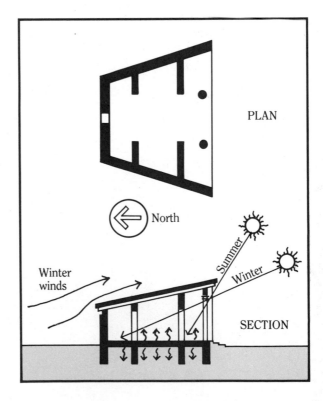

Figure 4-9 (right and left): Reconstruction of Socrates' house as described in Zenophon's Memorabilia.

PLAN

North

Winter winds

Summer

Winter

SECTION

Figure 4-10: Potential for passive solar heating in the United States.

used immediately, walls, floors, ceilings, and furniture store the excess heat, which radiates into the space throughout the day and night.

J. Douglas Balcomb and the research team at Los Alamos National Laboratory recommend that thermal mass be spread over the largest practical area in the direct-gain space. It is preferable to locate the mass in direct sunlight: Heat storage is as much as four times as effective when the mass is located so that the sun shines directly on it. The recommended ratio of mass surface to glass area is 6 to 1 (6 square feet of thermal mass surface for each square foot of window surface).

Locating thermal mass in interior partitions is more effective than putting it in exterior partitions, because heat can radiate from both sides of an interior wall. Thin mass is more effective than thick mass. The most effective thickness in masonry materials is the first 4

Figure 4-11: A direct-gain passive solar house. (Design by Dennis Holloway, architect.)

inches—thickness beyond 6 inches is pointless. The most effective thickness in wood is the first inch. Thermal mass should usually be a dark color to improve its heat absorption.

In northern climates, movable insulation in the form of insulating drapes, panels, shutters, or quilts often is used to cover the inside of the windows on winter nights to reduce heat loss. These items can also help ensure privacy, which may be at a premium since passersby can easily look through the same windows that the sun shines through. During the summer, the insulation can help block unwanted solar heat, although exterior shading devices (such as canvas awnings) will work better for this purpose than interior

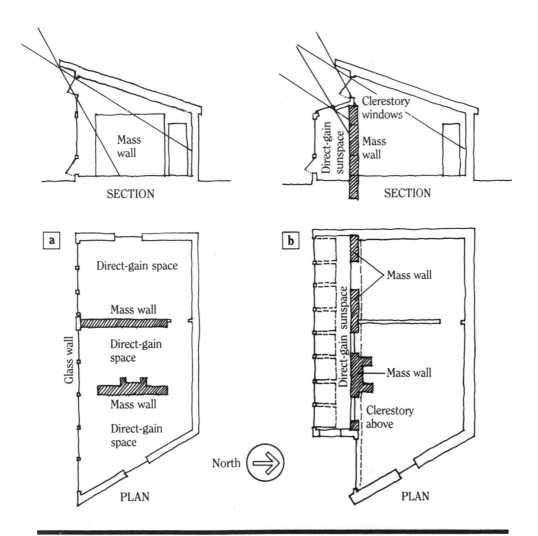

Figure 4-12: Internal mass storage walls serve as north-south partitions between direct-gain spaces (a) and as east-west partitions between direct-gain sunspace and north clerestory space (b).

devices. Deciduous trees and shrubs planted to cast shadows on solar-oriented glazing can shade the windows in summer. When the leaves drop, the winter sun can shine into the house.

Direct-Gain Sunspaces

A popular direct-gain heating strategy is the solar greenhouse or sunspace. Many homeowners claim this room becomes the favorite space in the house with its spacious outdoor feeling. The sunspace/greenhouse can, if properly designed and sited, provide as much as 50 percent of the house's heating requirements. In this situation, living spaces are better located on the south side with spaces not requiring as much heat (like bedrooms) on the north. Clerestory windows can be used in larger houses where it is important to get sunlight into the northside rooms.

If you plan to include a sunspace in your design, you'll first need to decide on the primary function of the space. Your primary goal may be to have a space to grow plants, or

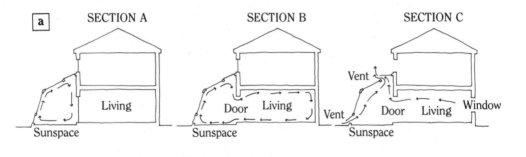

Figure 4-13a: One-story sunspace: winter, sunspace cut off from the house (Section A); winter, sunspace helps heat the lower story via open doors (Section B); summer, sunspace helps cool the lower story by pulling in air from the north windows (Section C).

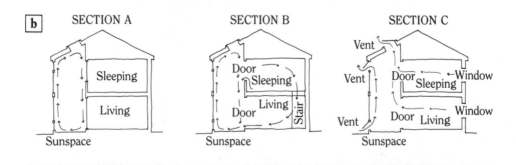

Figure 4-13b: Two-story sunspace: winter, sunspace cut off from the house (Section A); winter, sunspace helps heat both stories of the house (Section B); summer, sunspace helps cool both stories (Section C).

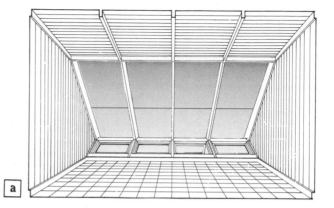

Figure 4-14: Sunspace with sloped south-wall glazing over reverse-slope vent windows (a). Sunspace with vertical south-wall glazing (sliding door), side venting windows, and sloped roof glazing (b). (Design by Dennis Holloway, architect.)

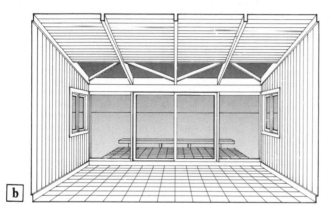

it may be to gather as much heat for the house as possible, or it may be to use the space as living space. It is possible to build a sunspace that will serve all three functions, but compromises will be necessary.

■ Growing plants: A sunspace intended primarily for plants will not provide much heat for the home. Much of the solar heat entering the greenhouse will be consumed by the plants themselves and the evaporation of water. One pound of evaporating water uses about 1,000 Btu of energy that would otherwise be available as heat.

Also, the sunspace will need to be ventilated. To stay healthy, plants need adequate ventilation, even in winter. There are air handling systems such as air-to-air heat exchangers that ventilate while retaining most of the heat in the air, but these add significantly to the cost of the project.

The light requirements of a space for growing plants call for overhead glazing, which complicates construction and maintenance, and glazed end walls, which are net heat losers.

■ Solar heat collector: If the purpose of the sunspace is to collect solar heat and distribute it effectively to the adjacent living space, you're faced with a different set of design criteria. Maximum gain is achieved with sloped glazing, few plants, and insulated, unglazed end walls.

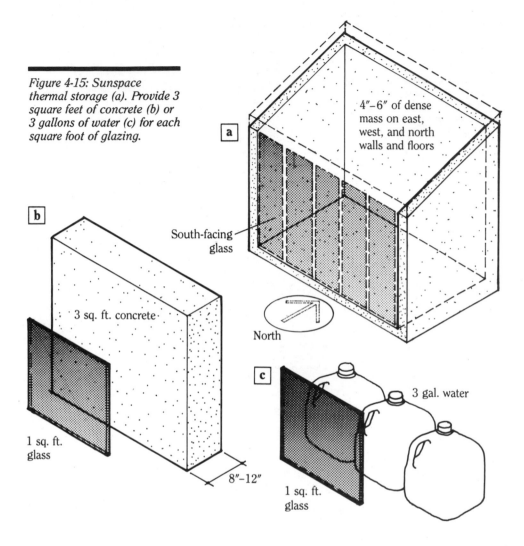

Figure 4-15: Sunspace thermal storage (a). Provide 3 square feet of concrete (b) or 3 gallons of water (c) for each square foot of glazing.

4"–6" of dense mass on east, west, and north walls and floors

South-facing glass

North

3 sq. ft. concrete

1 sq. ft. glass

8"–12"

3 gal. water

1 sq. ft. glass

You'll get more usable heat into your living space if there aren't plants and lots of mass soaking it up in the sunspace. Sun-warmed air can be moved into the house through doors or operable windows in the common wall, as well as blown through ductwork to more remote areas.

■ Living space: If your sunspace will be used as a room, you'll need to consider comfort and convenience in addition to energy efficiency. A room you plan to live in must stay warm in the winter and cool in the summer; it must have minimum window glare; and humidity must be moderate. It should also have carefully sized thermal mass to keep temperatures moderate both during the day and at night.

Vertical glazing may serve you best for this type of sunspace. First of all, although sloped glazing collects more heat in the winter, it also loses significantly more heat at night, which offsets the daytime gains. Sloped glazing can also cause a sunspace to overheat in warmer weather. Vertical glazing is cheaper and easier both to install and to insulate, and it is not as prone to leaking, fogging, breakage, and other glazing failures.

Indirect Gain

The second passive solar house type, indirect gain, collects and stores solar heat in one part of the house and then distributes this heat to the rest of the house. One of the more ingenious indirect-gain designs is called the Trombe wall. Named after its French inventor, the wall is constructed of high-density materials—masonry, stone, brick, adobe, or water-filled containers—placed 3 or 4 inches inside an expanse of south-facing glass. The wall is painted a dark color to more efficiently absorb solar radiation.

Heat collected and stored in the wall during the day slowly radiates into the house even up to 24 hours later. The Trombe wall allows efficient solar heating without the glare and ultraviolet light damage to fabrics and wood trim that is common in direct-gain solar houses. Trombe walls also afford privacy—passersby can't see through the masonry.

Some designers use "selective surface" materials, chrome-anodized copper or aluminum foils with adhesive backing, which can increase the absorptive efficiency of the wall to 90 percent, compared to 60 percent for a painted surface. These materials drastically reduce the amount of heat that the wall loses to the outdoors at night.

In several of the earliest Trombe-wall houses, small vents were used in the top and bottom of the wall: Heated air in the wall's air space would rise and pass through the upper

Figure 4-16: Trombe-wall house with attached sunspace, near Lyons, Colorado. (Design by Dennis Holloway, architect.)

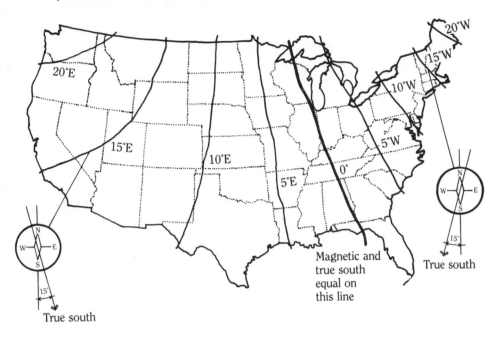

Figure 4-17: When designing a solar home, you must locate true (solar) south, not magnetic south. This map shows how magnetic south varies from true south in different parts of the United States.

vent into the house, while cooler room air would be drawn into the wall air space through the low vent. This is particularly effective in a building where heat is required quickly. However, the convective movement of air in the wall results in a significant decrease in efficiency over time due to the accumulation of dirt on the inner glass surface. Vented Trombe walls are known to be only about 5 percent more efficient, overall, than nonvented Trombe walls. Therefore, we generally recommend nonvented Trombe walls.

Designing a Passive Solar House

When the term "passive solar" was introduced in the 1970s, most people thought that if they wanted to build a passive solar house, they would have to hire not only an architect but a solar engineer capable of manipulating complex mathematical equations on a computer. Today, thanks to knowledge gained from a large number of completed "pioneer" passive solar houses, we are at a stage where even a high school student can design a passive solar structure.

In 1983, J. Douglas Balcomb and the research team at Los Alamos National Laboratory issued a set of design guidelines for heating passive solar houses. The five-step technique they devised gives owner-builders a solid basis for basic design decisions.

Step 1: Locate your building site on the map in figure 4-18. This will give you the conservation factor (CF) to be used in your house design. Note that for each geographic zone, the CF is expressed as a range. If your fuel costs are high, select the highest number.

Step 2: Use the following formulas to determine insulation values and recommended air-leakage rates for the house.

Wall R-values: Multiply the CF by 14. This is the R-value for the entire wall, including insulation, siding, interior sheathing, etc.

Ceiling R-values: Multiply the CF by 22. This is the R-value for the entire ceiling, including insulation, finish surface, etc.

R-value of rigid insulation placed on the perimeter of a slab foundation: Multiply CF by 13. Subtract 5 from this number. Use the same value for the insulation of the floor above a crawl space or for the perimeter insulation outside an exposed stem wall.

R-value of rigid insulation applied to the outside of the wall of a heated basement or bermed wall: Multiply CF by 16. Subtract 8 from this number. Use this value for insulation extending to 4 feet below grade. Use half this R-value from 4 feet below grade down to the footing.

Target ACH: Divide .42 by the CF. If the result is lower than 0.5 ACH, choose tight superinsulation techniques with controlled ventilation to maintain indoor air quality.

Layers of glazing on east, west, and north windows: Multiply the CF by 1.7, then choose the closest whole number. (If the number is 2.3, choose windows with two layers of glazing; if the number is 2.6, choose windows with three layers.) If the number exceeds 3, explore insulating glass and/or movable insulation.

Step 3: Next, compute a net load coefficient (NLC). To do this, look up your home's geometry factor (GF) in table 4-1. For example, if the house will have a total floor area of nearly 3,000 square feet on three stories, the GF will be 5.7.

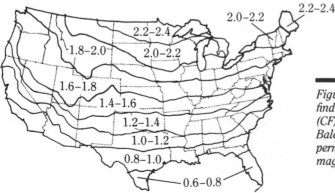

Figure 4-18: Use this map to find your conservation factor (CF). (Source: J. Douglas Balcomb, et al. Redrawn with permission from Solar Age magazine.)

Table 4-1 **GEOMETRY FACTOR (GF)**

Floor Area	Number of Stories			
(sq. ft.)	1	2	3	4
1,000	7.3	–	–	–
1,500	6.5	6.7	–	–
3,000	5.4	5.4	5.7	–
5,000	4.9	4.7	4.9	5.1
10,000	4.3	4.0	4.0	4.2

Now multiply the GF by your house's floor area. Thus, if the floor area will be 2,900 square feet and the GF is 5.7, you multiply these two values to get 16,530. Finally, divide this result by the CF. If your CF is 2.0, for example, you would divide 16,530 by 2 to get 8,265. This is your NLC.

Step 4: Locate your building site on the map shown in figure 4-19. This will give you the load collector ratio (LCR) for your home. Note that for each geographic zone, the LCR is expressed as a range. If your fuel costs are high, select the *lowest* number.

Step 5: To determine the area of a passive solar collector (Trombe wall, sunspace, etc.) for your home, divide the NLC (the number you got in Step 3) by the LCR (the number you got in Step 4). For example, if your NLC is 8,265 and your LCR is 20, then your passive solar collector should have 413 square feet of south-facing glazing. You can round this number up or down by 10 percent (so the area could be as small as 370 square feet or as large as 450 square feet). In hot climates, the area should be adjusted downward by 20 to 30 percent.

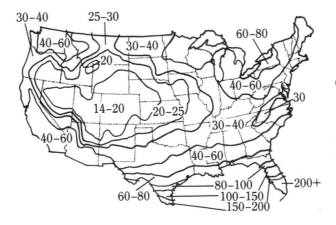

Figure 4-19: Use this map to find your load collector ratio (LCR). (Source: J. Douglas Balcomb, et al. Redrawn with permission from Solar Age *magazine.)*

UNCONVENTIONAL CONSTRUCTION METHODS

One of the advantages of building your own home is that you can declare your freedom from conventional construction techniques. In this section of the book, we will describe some techniques that may be unusual in your neighborhood but that you may want to consider. They can enable you to create a home that is both unique and well adapted to your personal needs.

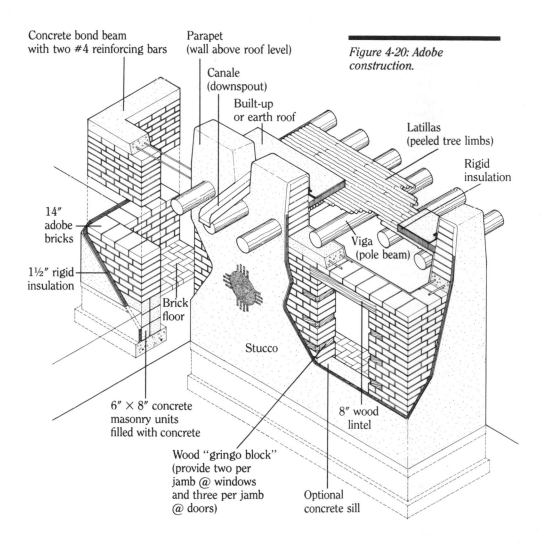

Concrete bond beam with two #4 reinforcing bars

Parapet (wall above roof level)

Canale (downspout)

Built-up or earth roof

Figure 4-20: Adobe construction.

Latillas (peeled tree limbs)

Rigid insulation

14" adobe bricks

1½" rigid insulation

Viga (pole beam)

Brick floor

Stucco

6" × 8" concrete masonry units filled with concrete

8" wood lintel

Wood "gringo block" (provide two per jamb @ windows and three per jamb @ doors)

Optional concrete sill

PERSPECTIVE

*Figure 4-21: Right and left):
This passive solar adobe
home was built using the
poured adobe method.
(Adapted with permission
from drawings by Mike
Belshaw.)*

Adobe

Earth is very likely the oldest building material known to humanity, and it is still widely used today. The strength, durability, and availability of earth makes it particularly attractive when other building material resources are in short supply. The "adobe style" has been energetically revived in the Southwest, for example. Some of the modern adobe homes in New Mexico, Arizona, and Colorado are among the most luxurious homes to be found anywhere.

Traditional "puddled" adobe bricks are produced from soil, ideally taken from a location as close to the building site as possible. The soil is mixed with water, shaped in a mold, dried in the sun, and then used to build walls. The earth mix is fairly sloppy, about the consistency of very heavy cream, hence the term "puddled." The mortar used has the same composition as the bricks, except that the larger stones are screened out of it. The bricks must be cured before they are laid onto the wall; curing time will vary with humidity and temperature, but takes a minimum of a couple of weeks.

A variation of the traditional technique involves the use of a hydraulic ramming device to press the mud into bricks. Rammed adobe bricks can be produced at the astonishing rate of one every 12 seconds. The earth used to make these bricks has a relatively low moisture content, around 10 or 11 percent, so they require a minimum of curing and can often be taken off the machine and built into a wall. They are more costly than traditional adobe bricks, but often the time and labor saved justifies the extra expense.

A major disadvantage to building with adobe bricks is that the technique is extremely labor-intensive. "Poured adobe" is a building technique that helps solve this problem. The best-known proponent of poured adobe is Mike Belshaw, who built two houses using the technique on a ranch he owned near Prescott, Arizona. His first house (1,500 square feet), built in 1975, cost $20 per square foot; the second house (4,000 square feet) was completed for around $30 per square foot.

Specially designed portable forms are the heart of the system. The first type of form is a collapsible box made with two 1" × 12" × 36" boards and two 1" × 12" × 24" boards (see figure 4-22). The box is hinged on one end and held together at the other with ⅜-inch all-thread rods. The plate at the unhinged end can be canted to create bricks that are thicker on the bottom than at the top, thus permitting the construction of tapering walls in the style of old adobe buildings. Several of these forms are built and placed on the wall or footing about 24 inches apart.

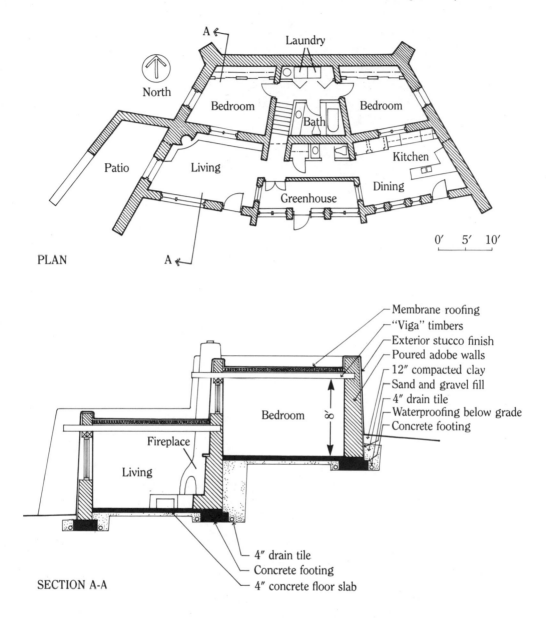

PLAN

SECTION A-A

Membrane roofing
"Viga" timbers
Exterior stucco finish
Poured adobe walls
12" compacted clay
Sand and gravel fill
4" drain tile
Waterproofing below grade
Concrete footing

4" drain tile
Concrete footing
4" concrete floor slab

Dry adobe soil is loaded into a cement mixer or mixing truck through a 4-inch mesh, and water is added in the correct proportion. The adobe mud is then placed into the forms and tamped down. Depending on the temperature and humidity, the forms can be removed in 4 to 24 hours. The forms are then moved to a new section of wall where the process is repeated. The spaces between the resulting adobe bricks are filled by attaching two plates made of 1" × 12" × 36" boards wired together (see figure 4-22). Adobe mud is placed between these boards.

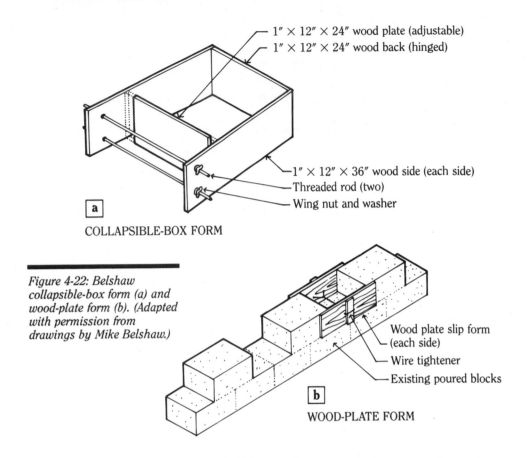

1" × 12" × 24" wood plate (adjustable)
1" × 12" × 24" wood back (hinged)

1" × 12" × 36" wood side (each side)
Threaded rod (two)
Wing nut and washer

a

COLLAPSIBLE-BOX FORM

Figure 4-22: Belshaw collapsible-box form (a) and wood-plate form (b). (Adapted with permission from drawings by Mike Belshaw.)

Wood plate slip form (each side)
Wire tightener
Existing poured blocks

b

WOOD-PLATE FORM

Poured adobe is a fast way to build an earthen house. Belshaw says that a three-person crew, working hard on a simple structure, can get a wall to roof height in about 16 working days. The greatest obstacle to the method is that it is not covered by model building codes, so you may need to work hard to win approval from building officials and lenders.

Rammed Earth

Rammed earth, also known as pise de terre or pise, is a method of constructing earthen walls by building up layers of compacted earth. Wooden forms are placed on a footing, and earth is placed in the form and tamped down (or "rammed") in 4- to 6-inch layers until the forms are full. The forms are then moved up, new soil is put in and rammed—then the forms are moved upward again. The process is repeated until the wall reaches the desired height.

A rammed earth wall is self-supporting as soon as the ramming is complete, although it can take several years to reach full strength. After about two days, the wall is cured enough to install doors and windows. If you wait more than a couple of weeks to do this work, the wall may be so hard that you will no longer be able to nail into it and will have to drill with a masonry bit before fastening framing to it.

Owner-builders can build their own rammed earth walls by hand or hire a contractor (if such a person exists in your area) to build only the walls then do the rest of the work on the house themselves at a considerable cost savings. A 16-inch-thick wall may cost as little as $4 per square foot, while a 36-inch-thick wall may cost about $8 per square foot. These costs include the "bond beam," a code-required reinforced concrete beam that runs along the top of an 8-foot-high wall, or at the top of each story in a two-story structure. Footing costs are not included.

Soil Mixtures

Soil suitable for earth construction is found nearly everywhere in the world. Ideally, the soil should contain 70 percent sand and aggregate and 30 percent clay mix. Variations from these proportions may still result in strong walls, as long as the clay percentage is kept low. If the proportion of sand and aggregate is increased, the resulting wall will be less resistant to erosion but will have greater compressive strength. The "bottom line" test for any soil is to make a few test blocks and see how strong and water resistant they are.

It is a common myth that earth construction is best used in hot, arid climates such as New Mexico. While it is clear that hot, arid climate zones are the primary locations for earthen buildings, it is also clear that earth works as a building material in wetter northern and eastern climates. Rammed earth walls do not require the curing time of adobe bricks, so they are especially well suited to wet, humid climates, where curing bricks would be difficult.

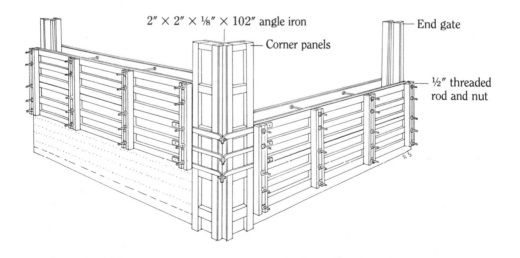

Figure 4-23: A forming system for rammed-earth walls, developed by David and Lydia Miller of Greeley, Colorado. (Redrawn with permission from Rammed Earth Institute International.)

Some builders "stabilize" their earthen walls to increase the weather resistance of the material. Traditional earth builders used straw, animal manure, and blood as stabilizing agents, but Portland cement or asphalt emulsion are more common today. Stabilizing eliminates the need for extra-long overhanging eaves, which might block the sun in a solar building.

Energy codes, where applicable, will require that an earth wall be insulated, since earth doesn't have a high R-value. It makes the most sense to insulate on the exterior with one of the foam-and-stucco systems such as Dryvit (Box 1014, 1 Energy Way, West Warwick, RI 02983). This keeps the walls on the inside of the insulation where they can function as thermal mass to moderate indoor temperatures. If you use such a system, there is little point in going to the expense of stabilizing the earth, since it will never be exposed to weather.

The Owner-Built Earth House

The main technical advantage of earth construction for owner-builders is that there are few "exact" rules that must be followed. There are rules, to be sure, but they are not as rigid as those for frame construction or even log building. If you're willing to put in the time to gather the information, earth construction is easy to learn.

Earth construction has low materials costs and high labor costs (if you hire workers), so there is a real advantage to owner-builders who have the time and skill to do most of the labor themselves. A crew of two can produce 300 to 800 bricks per day, depending on the equipment they use. After the bricks have cured, a team of three can lay up as many as 800 bricks per day while standing on the ground and about half that number while working on a scaffold. Some owner-builders say they spent $10,000 to $30,000 on adobe homes that have market values of $80,000 to $150,000.

If your local code does not recognize earth construction, write to Construction Industries Division, Bataan Memorial Building, Santa Fe, New Mexico for a copy of the New Mexico Adobe Code, Section 2405, Amendment 6: Chapter 24 Masonry. Make sure you get the current amendment. The New Mexico code should be a valuable aid in convincing your building department that earth structures are safe and structurally sound.

Earth-Sheltered Houses

Earth-sheltered houses, sometimes referred to as "underground" houses, can be defined as homes that are built into or surrounded by earth. Energy conservation and protection from storms are major incentives for most owners of earth-sheltered houses. The decision to build such a house will profoundly affect your choice of a building site, the materials you use, and the design and layout of the living space.

Gideon Golany, an authority on earth-sheltered homes, lists the following climates as being suitable for subterranean construction: (1) very warm and dry, such as the southwestern United States; (2) very cold and dry, such as central Canada; (3) temperate (cold and snowy in winter, rainy and relatively warm in summer), such as the upper midwestern United States.

Although other houses, such as superinsulated and passive solar homes, achieve similar levels of energy efficiency, the earth-sheltered house has several unique advantages:

■ Temperatures below the surface of the ground are much more stable than air temperatures. In Minnesota, for example, the air temperature fluctuates yearly as much as 130°F (from −30°F to 100°F), while the temperature of the earth 10½ feet below the surface fluctuates only about 18°F (from 40°F to 58°F)—a mild underground environment compared to the extreme aboveground environment.

■ The earth gives up its summer heat slowly in the autumn and warms up slowly in the spring. This "thermal lag" means that the heating and cooling seasons are postponed for an underground house. If the soil outside your walls remains warm in the autumn, you won't need as much heat as you would if your house stood aboveground, exposed to chilling winds. Similarly, you won't need to cool your house in the spring or early summer if the soil outside your walls stays cool.

■ An earth-sheltered house is more likely to withstand the effects of severe storms such as tornadoes. The house hunkers down, where the winds can't hurt it.

■ If concrete and concrete masonry are the primary building materials, the house will be more rot-proof, vermin-proof, and fire-resistant than an above-grade house built with conventional materials.

■ The sound-dampening effects of earth make earth-sheltered buildings much quieter than above-grade structures.

■ Malcolm Wells, the well-known architect and proponent of underground architecture, says that his main reason for going underground is "because it is so beautiful."

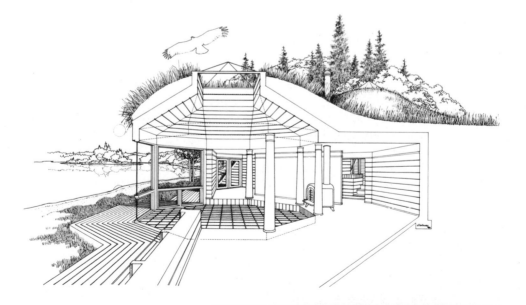

Figure 4-24: Underground homes can be beautiful as well as practical. (Design by Dennis Holloway, architect.)

Site and Structure

Sites with expansive soils or with bedrock or water close to the surface are not suitable for homes built under the surface. Nearly any site problem can be solved, but the solutions can be very costly and/or disruptive to the local environment. For instance, bedrock can be blasted away, but this is not a do-it-yourself job. A better solution would be to build aboveground and berm (pile earth against) the walls. If there isn't enough soil on the site for this, you'll have to haul it in, which can be expensive. It is always better to design your house for the site, rather than forcing an inappropriate structure onto a site.

If you know you want to build an earth-sheltered home and don't own the land yet, look for property with soil that drains well, that can be excavated easily, and that can easily carry the load of your home. Obtaining a soils report before you buy is good insurance against expensive or disappointing surprises.

Earth-sheltered buildings are subjected to heavier loads than above-grade structures, since earth presses against the walls and (usually) the roof. (Some earth-sheltered homes have conventional roofs that protrude above ground level.) Considerable attention must be paid to adequate structural engineering. It is advisable to consult a professional structural engineer familiar with earth-sheltered design at the outset of the design process. Fortunately, there are a variety of materials to choose from to carry the loads borne by an earth-sheltered home.

Poured concrete or concrete block are logical choices for earth-sheltered construction, since concrete and masonry products are rot- and vermin-proof and have high compressive strengths. If you plan to do most of the work yourself, concrete block is a better choice than poured concrete, since laying up block is far easier than orchestrating massive concrete pours. We would suggest either a surface-bonded block or interlocking block wall (see chapter 5) rather than conventional mortar-jointed block. These systems are easier to build, easier to waterproof, and stronger than the more conventional mortared wall.

Another possibility for the dedicated do-it-yourselfer is all-weather wood (i.e., wood impregnated with wood preservative chemicals). We have reservations about the toxicity of the preservatives used in the wood, but the fact that you could build your underground house using basic framing techniques make all-weather wood an attractive option for an owner-builder. See chapter 5 for a discussion of all-weather wood foundations.

Rob Roy has done some interesting work with low-cost "cordwood masonry" houses in the severe climate of upstate New York. We will describe this technique near the end of this chapter. Roy's Earthwood Building School offers instruction in the technique (see Appendix B).

Other materials such as thin-shell concrete domes and precast concrete planks are not accessible to the owner-builder, but it is possible to hire professionals to erect the basic structure, and then you do all the finishing work yourself. Precast concrete is a commercial building material—you can probably find suppliers in your area by checking the yellow pages or calling commercial contractors. Thin-shell concrete domes must be built by contractors with the proper equipment and experience.

Insulation

Despite the energy advantages of earth-sheltered construction, these houses must be insulated. There are currently many design theories about the amount and placement of insulation around an earth-sheltered house, but most experts agree on the following points: (1) insulating below the floor slab is not cost-effective; (2) where walls are not covered by earth or earth berms, insulating the perimeter of the foundation walls and insulating around the edge of the slab are good investments; (3) if the house is built of masonry, insulation should always be placed outside of the walls, which allows the thermal mass of the walls to moderate temperatures in the living space. Exterior insulation will also protect the waterproofing from abrasion during backfilling and from seasonal movement caused by freeze/thaw cycles.

In a study reported by the National Concrete Masonry Association, *Energy Savings through Earth Sheltering* (TEK 119), four cases for insulation placement were evaluated:

Case 1: If there is less than 5 feet of soil on the roof, the ceiling should be insulated. This is more cost-effective than to increase the structural strength of the house in order to carry a deeper earth covering.

Case 2: Insulating the ceiling and walls down to 6 inches below the frost line results in medium heat loss during both winter and summer and is well suited for temperate climates requiring both heating and cooling.

Case 3: Insulating the ceiling and the walls down to the footing yields low heat loss. This case is best suited to northern and high-altitude climates where heating requirements dominate the year.

Case 4: A good strategy in most climates is to place insulation on the roof, partially backfill the walls, place rigid closed-cell insulation outside this earth, and then continue backfilling over the insulation. The amount of earth in contact with the walls is thereby increased while the frost line is lowered. The earth between the walls and the insulation is shielded from the aboveground temperatures, resulting in excellent building performance in both winter and summer.

Figure 4-25: Four cases for insulation placement.

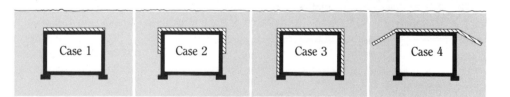

Drainage and Waterproofing

If water leaks from the soil into an earth-sheltered home, you won't have just a wet basement—you'll have a wet living room, kitchen, and bedrooms. So preventing such leaks is critical. We recommend overdoing your waterproofing and drainage system to assure that you will never find yourself digging up your house to repair a leak.

A relatively inexpensive waterproofing system might include a layer of roofing felt, roofing cement troweled on over that, polyethylene sheeting (6 mil or greater) with generous overlaps over the cement, topped with more roofing cement, sand, and sod. Unfortunately, given the manufacturing defects and degradation that polyethylene is prone to, this system is probably a little risky.

A "built-up" roof may also be a relatively inexpensive answer, although there is some debate about how advisable such a roof is in a situation where it can't be easily inspected and repaired. The roof incorporates multiple layers of roofing felt and asphalt or pitch. Glass-fiber membranes rather than organic felts are recommended, since the latter will rot if they are exposed to water over a period of time.

More sophisticated coatings are also available. Bentonite clay is very expansive when moistened, and as the molecules expand in a confined space, like on a bermed wall, they press closely enough against each other so that no water can pass through. There are several methods of applying this material, but it is generally not a do-it-yourself project. The warranty of at least one product, Bentonizing, is void unless the product is applied by a licensed applicator. Bentonite clay requires moisture to do its job, so it is not recommended for very arid climates unless precautions are taken to assure that it will never dry out.

A number of other products including butyl, EPDM, and neoprene membranes are also available. Although they are among the most expensive choices, remember that this is not a job you want to have to do again.

A successful system combines an effective waterproofing treatment with a drainage system that never allows water near the wall. Drain tile set in gravel must be set around the footing of the underground wall, and another draintile at the top of a totally buried wall is also a good idea.

Earth on the Roof

The roof of an underground house can be covered with sod or several feet of earth. This can be an advantage in both summer and winter. Although earth is not a good insulator, it can be used to enhance the thermal effectiveness of a well-insulated roof. In very cold climates, sod can help to maintain the accumulation of snow on the roof— especially north-sloping or shaded roofs—which further reduces heat and infiltration losses. The cooling benefits in the summer are even clearer. Evaporation of rainwater and morning dew and the transpiration of moisture through grass blades can reduce the temperatures of the roof by 20°F.

The sod roof often becomes a bed for wildflowers. Earth several feet thick covering the roof of an underground house makes the growth of root structures of larger shrubs possible, producing a lush hanging garden effect. Planting indigenous plant species on the roof and surrounding earth berms can result in the illusion that the house has always been there.

As you might imagine, an earth-covered roof can create severe waterproofing and structural problems if you do not do your homework or hire experts with earth-sheltering

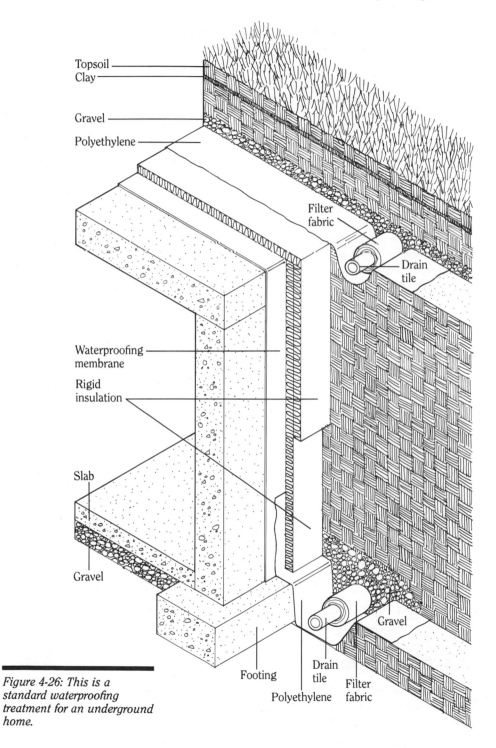

Topsoil
Clay
Gravel
Polyethylene
Filter
fabric
Drain
tile
Waterproofing
membrane
Rigid
insulation
Slab
Gravel
Gravel
Footing
Drain
tile
Polyethylene
Filter
fabric

*Figure 4-26: This is a
standard waterproofing
treatment for an underground
home.*

experience to help you with the design. Slightly sloping the roof is an inexpensive strategy that helps minimize leaks. Exposed flat roofs are notorious leakers—just ask any construction law attorney!

Codes and Financing

Like any other home that uses an unconventional building method, your earth-sheltered home is apt to run into trouble with your local building department and your lender. The building department will require a carefully engineered structural design before they issue the building permit, and building officials are likely to watch the construction process very carefully. Your best defense against problems is to develop a good working relationship with the officials early in the process. Find out what will be required of you before you apply for the building permit, to save costly and frustrating delays.

Your lender's major concern is the marketability of the property. Lenders are hesitant to lend on unconventional houses because experience has taught them that the market for such houses is much narrower than for a more conventional house. If you should default on your loan, the lender is left with a property to sell that appeals to only a small minority of the home-buying public.

The burden is on you to educate the individuals and organizations that have the power to veto your project. Your architect (assuming you use one) and structural engineer will be powerful allies in this process. And you should educate yourself thoroughly about the type of house you plan to build. The time you spend doing so will not be wasted—having to satisfy the building department and lender will sharpen your own thinking about the project.

Costs

The cost of an earth-sheltered home depends on a myriad of variables. It is possible to build an underground home for less than you would spend on a comparable aboveground home, if you use inexpensive materials and do much of the work yourself. But the cost of excavation, a job that typically gets hired out by all but the most fanatical owner-builders, is apt to be higher than for a similar conventional home, and the extra structural strength required to retain earth and support earth roofs can result in construction costs 30 percent higher than for aboveground construction.

The real savings in an earth shelter will be "life cycle" savings—what it costs to live in the home over the life of the structure. Heating and cooling costs should be low, and almost no exterior maintenance will be required. For those of us who don't relish the prospect of painting or staining wood siding every few years, this is a major attraction.

Masonry Houses

For years, concrete masonry has been used as a moderate-cost alternative construction technique, with components constantly being improved. While in the past concrete masonry was used primarily for foundation walls, the benefits of using masonry for the entire home have become widely appreciated in recent years. As we mentioned earlier, increasing the amount of thermal mass in a home can have distinct benefits during the heating season, especially in passive solar applications: The mass stores heat during the day and releases it slowly during the night. Concrete masonry, like adobe and poured concrete, is an excellent thermal mass.

Figure 4-27, taken from the *Passive Solar Design Handbook,* by J. Douglas Balcomb, et al. (1980, National Technical Information Service, U.S. Department of Commerce, Springfield, Va.), shows the "diurnal heat capacity" (the daily heat stored and returned by various building materials in a direct-gain passive solar building). As you can see, concrete block has the best heat-storage capacity, followed by brick. The graph assumes that the storage materials are backed by insulation or by an equal thickness of the same storage material that is also used to store solar heat.

Besides its benefits during the heating season, concrete masonry can result in significant savings during the cooling season. It is now generally recognized that a well-insulated masonry home may not require expensive mechanical air-conditioning in climates where a conventional frame house would be uncomfortable without it. This can mean a considerable savings to the owner-builder in that the money for air-conditioning equipment can be put to other uses. The house will also be cheaper to live in, since cooling costs will be minimal or nonexistent.

In very warm climates, the thermal mass of solid-grouted masonry walls reduces the transfer of unwanted heat from the outside to the inside of the building and also absorbs unwanted heat that may have accumulated on the interior of the building. Best results are

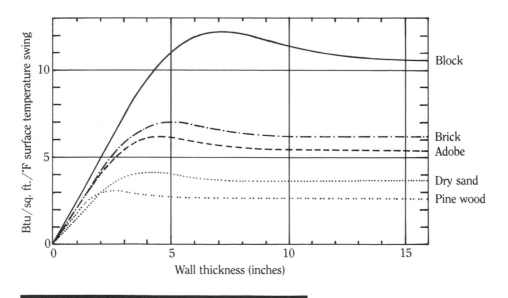

Figure 4-27: Daily heat stored and returned for different materials. (Source: Passive Solar Design Handbook, *by J. Douglas Balcomb, et al.)*

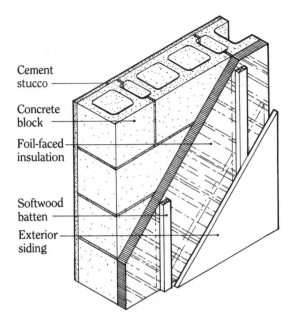

Cement
stucco

Concrete
block

Foil-faced
insulation

Softwood
batten

Exterior
siding

Figure 4-28: Recommended technique for insulating masonry walls: radiant wall barrier system. (Adapted with permission from the National Concrete Masonry Association, Concrete Masonry for Natural Cooling-*TEK 139.)*

obtained by shading or earth-sheltering these walls, especially on the west side of the house. Placing the garage on the west is an effective strategy for reducing afternoon solar heat gain.

The wall itself can reduce heat transfer if it is constructed in the following manner, as recommended by the National Concrete Masonry Association (NCMA): First, fix foil-faced insulation or double-sided builders' paper over the outside of the block; next, apply wood battens vertically and cover with the exterior siding, making sure to vent the enclosed air spaces. The outside of the concrete block is kept much cooler this way, allowing the whole concrete wall to absorb unwanted interior heat more effectively.

Figure 4-29: Recommended exterior insulation thickness based on R-5 per inch nominal rigid insulation. (Adapted with permission from the National Concrete Masonry Association, Exterior Insulation of Block Walls-*TEK 134.)*

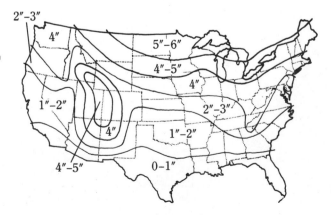

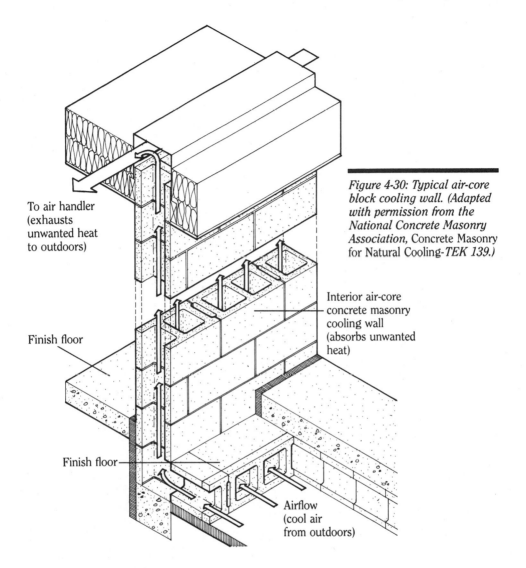

To air handler
(exhausts
unwanted heat
to outdoors)

Figure 4-30: Typical air-core block cooling wall. (Adapted with permission from the National Concrete Masonry Association, Concrete Masonry for Natural Cooling-*TEK 139.)*

Finish floor

Interior air-core
concrete masonry
cooling wall
(absorbs unwanted
heat)

Finish floor

Airflow
(cool air
from outdoors)

Another technique recommended by NCMA involves the use of hollow-core masonry walls inside the house as thermal mass. These walls can be cooled by nighttime outside air driven by fans through the cores, thus cooling the house's interior. This "air-core" wall concept is also effective in controlling winter indoor temperatures by recycling warm air through the cores.

Log Homes

Building homes with logs is a time-honored tradition. If a rustic aesthetic appeals to you, there are few building methods that offer the sense of comfort and security that a log

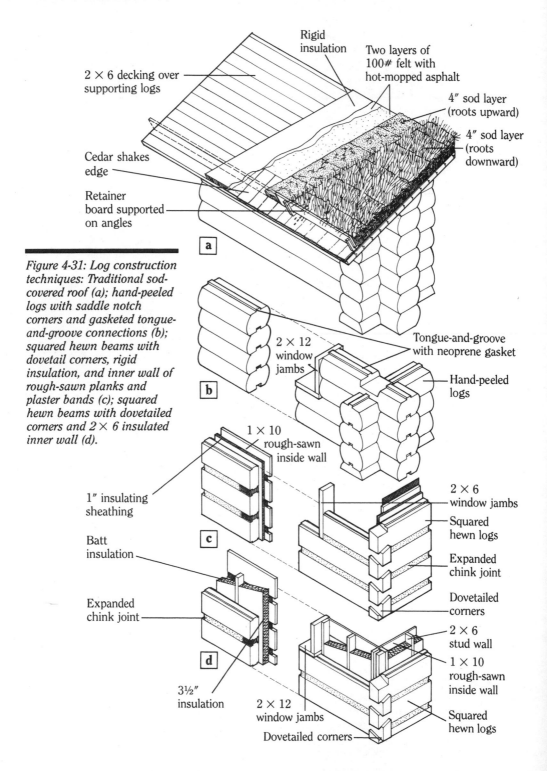

2 × 6 decking over supporting logs

Rigid insulation

Two layers of 100# felt with hot-mopped asphalt

4" sod layer (roots upward)

4" sod layer (roots downward)

Cedar shakes edge

Retainer board supported on angles

a

Figure 4-31: Log construction techniques: Traditional sod-covered roof (a); hand-peeled logs with saddle notch corners and gasketed tongue-and-groove connections (b); squared hewn beams with dovetail corners, rigid insulation, and inner wall of rough-sawn planks and plaster bands (c); squared hewn beams with dovetailed corners and 2 × 6 insulated inner wall (d).

2 × 12 window jambs

Tongue-and-groove with neoprene gasket

Hand-peeled logs

b

1 × 10 rough-sawn inside wall

1" insulating sheathing

2 × 6 window jambs

Squared hewn logs

Expanded chink joint

Dovetailed corners

c

Batt insulation

Expanded chink joint

2 × 6 stud wall

1 × 10 rough-sawn inside wall

d

3½" insulation

2 × 12 window jambs

Squared hewn logs

Dovetailed corners

home does. Typically, the skills needed to erect a log home are within reach of a motivated novice.

For the owner-builder who wants to build a log home, there are essentially two choices: purchase a kit or cut your own logs and do your own joinery. Over 150 companies make log home kits in the United States. Some of the kits fit together as readily as a child's Lincoln Log set.

Air leakage is an especially vital concern in a log home, since every seam between the logs is a potential leak, and there are lots of seams. Also, logs are almost certain to shrink over time, and they will tend to expand and contract with changes of humidity and temperature. The intersections between the logs should be sealed with a material that can handle these changes, or you'll find yourself having to renew the seal on a regular basis.

Log home manufacturers have made great strides over the years in developing effective sealing techniques. Comparing the techniques offered by various manufacturers should be high on your list of priorities as you shop for a kit. There are foam gaskets, flexible chinking, and caulks that do the job if installed properly. Pay special attention to the sealing at intersections between the log walls and the foundation, the log walls and the roof structure, and around any openings in the walls.

Energy Efficiency

There is some controversy surrounding the relative energy efficiency of log structures. Much of the debate revolves around the R-value of a solid wood wall. Uninsulated log walls don't satisfy the minimum R-value requirements of most energy codes. Different species of wood have different R-values, ranging from about .71 per inch for shagbark hickory to 1.41 for northern white cedar.

R-values, however, tell only part of the overall energy-efficiency story. The type and placement of windows, orientation of the building, airtightness of the structure, amount of thermal mass, levels of insulation in other parts of the house, and the climate where the

Figure 4-32: The structure of the end of an old log.

Figure 4-33: This log cottage near Hjeltar, Norway (1760 A.D.), combines vertical and horizontal logs. (Adapted from Early Wooden Architecture in Norway, *by Gunnar Bugge and Christian Norberg-Schulz.)*

house will be built must all be considered to get a picture of how the house will function in the real world.

Log home manufacturers have long insisted that the low R-value of wood is offset by the massiveness of the logs themselves. They claim that the thermal mass effect is significant enough that log homes show overall energy performance equal to or exceeding that of conventionally insulated frame walls. And indeed a study conducted in Maryland showed that a log home may well outperform a conventional home in all but the harshest months of the winter. But bear in mind that Maryland has a mild climate; in general, we believe there are easier and more cost-effective ways to build an energy-efficient home in a severe northerly climate than with logs.

If you have a strong preference for log homes and live in a severe climate, we would strongly suggest that you have the design computer modeled for energy efficiency to assure that the structure will be as efficient as possible. Your log home dealer, local energy extension office, local owner-builder organization, or architect should be able to put you in touch with someone in your area who offers this service. If not, there are organizations around the country that can use data from your area to help you refine the energy efficiency of your log home. One good contact is the Log Home Guide Information Center, Exit 447, Interstate 40, Hartford, TN 37753.

Domes

The strength and durability of dome structures is well documented. While domes have been built of a variety of materials, including brick, concrete, and foam, the technique most accessible to today's owner-builder is the wood geodesic dome.

A geodesic dome is a system of triangles that, when fitted together, approximate a half sphere. Since the triangle is the only self-bracing construction form, the resulting structure is extremely strong. As the number of triangles used increases, a closer approximation of the spherical form is achieved, and strength increases because the surface stress is more efficiently distributed.

In addition to being exceptionally strong, the dome encloses the maximum amount of space with the minimum amount of materials. It exposes the least possible amount of surface area to the elements and thus is economical to heat and cool. Like conventional frame structures, geodesics can be developed as passive solar homes, with direct-gain windows, sunspaces, and thermal walls and floors.

In traditional barn-raising fashion, many owner-builders have erected the shells of large dome homes in one or two days with a crew of family and friends—without any professional help. The money saved through the inherent economies of the building method can make a geodesic dome one of the more cost-effective building systems. An estimated 5 to 15 percent savings over other types of frame housing is possible because a geodesic dome uses about 30 percent less material to enclose the same square footage. Because of the amount of angular cutting involved, domes take longer to finish than a conventional home does; however, because a finite set of angles are symmetrically repeated, careful measuring and cutting will minimize waste of time and materials. Dome kits can greatly save labor time, but you will be paying a premium.

Geodesics do have a few drawbacks. Because they consist of triangles, cutting and fitting of insulation materials can be time-consuming. Spray foam and cellulose insulation are more adaptable to this building style, but they are more expensive and require special equipment and professional skills. A related problem results from the entire wood structural frame acting as a "thermal bridge," conducting heat from the inside to the outside of

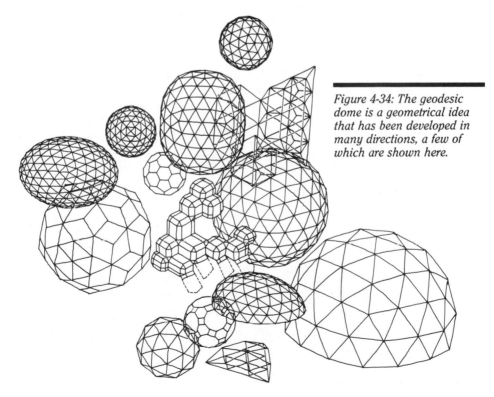

Figure 4-34: The geodesic dome is a geometrical idea that has been developed in many directions, a few of which are shown here.

the building skin—a problem which in conventional construction can be solved by applying exterior insulation over the sheathing.

Whether you design your own dome, purchase a set of plans, or buy a kit, always make sure that a professional engineer approves of the stuctural design for the struts and the connector hubs before you begin.

Concrete-and-Foam Domes

If you're drawn to the idea and aesthetic of living in a dome home, but are discouraged by the drawbacks of geodesics, consider a thin-shell concrete dome (also called "monolithic domes"). Because the process requires specialized equipment and skill, building a concrete dome is not a do-it-yourself project; however, it is possible to do your own finish work.

Building the concrete shell involves five basic steps:

1. After the foundation is poured, an air bag the size and shape of the desired structure is attached to the foundation and inflated. Urethane foam insulation is sprayed on the inside of the balloon to a predetermined depth, usually about 4 inches.

2. Openings for doors, windows, and other openings are prepared.

3. In some cases, the next step is to place pre-engineered steel reinforcement next to the foam. Some builders use an engineered concrete mix reinforced with needle-sized bits of steel instead.

4. Reinforced either way, the interior of the foam dome is then sprayed with 1½ to 4 inches of concrete. When the concrete cures, you have a concrete dome with exterior insulation. Among other advantages is the fact that the concrete serves as fireproofing for the foam.

5. The final step in completing the shell is to coat the outside surface of the foam. Urethane foam decomposes in sunlight. Several types of coatings are available to protect the foam. Your choice should be based upon color, texture, thickness, cost, and durability (the coating's ability to filter out ultraviolet radiation, to expand and contract in response to temperature changes, and to breathe). Outgassing from the foam will cause bubbles to form in the coating unless the gases can escape.

Good Neighbors

Some areas specifically forbid dome buildings. Before you buy property, check the covenants of the subdivision and the zoning regulations of the local government that has jurisdiction over the lot you plan to buy. Even if there isn't any official opposition, you may want to run your plans by your future neighbors to make sure you're not making enemies before you even move in. Domes have an aesthetic that people either love or hate, so you may want to preserve any mature trees on the property to help blend the dome into its surroundings.

It seems likely to us that domes will become increasingly popular, particularly as labor, materials, and energy costs continue to rise. As more domes are built, more people will become more comfortable with the form. Although we realize the sample is self-selecting, it seems worth noting that virtually every person we spoke to who lives in a dome was passionately enthusiastic about his home. This enthusiasm is sure to be contagious.

Timber Framing

Timber framing is a technique that goes back at least 2,000 years, and there are many timber frame structures still in use after 800 years. Before the development of nails and joinery hardware, timber framing was about the only way to build with wood—the frames were held together with wooden pegs. As nails came into more common use, timber framing went into a decline, largely because of the skill and care required to successfully use the technique. But timber framing is now enjoying a revival, and the combination of time-tested joinery techniques and modern tools is bringing these homes within financial reach of the average homeowner. Timber framing is a particular favorite among owner-builders because of the exceptional energy efficiency, beauty, and structural integrity of these buildings.

In contrast to conventional wood framing, timber framing uses large beams resting on large posts rather than 2-inch dimensional lumber to support the building. The posts and beams can be spaced far apart, which allows for larger openings in the exterior walls and increases your design flexibility on the interior of the home, since interior walls can be placed without concern about structural considerations.

The posts and beams can be left exposed on the interior, which adds warmth and drama to the home. The timbers are often planed and finished with oil, and the massiveness of these large wood members creates a sense of integrity and security that is impossible to duplicate in a conventionally framed home. Some talented carpenters carve the timbers, truly customizing the home.

Most framers consider oak to be the wood of choice for a timber frame. The timbers are connected by cutting precise joints, carefully fitting these joints, and fastening them together with hardwood pegs. Timber framers prefer pegs over metal fasteners for a couple of reasons. First, as the frame dries and settles, a metal connector will cause the joint to loosen. Wood and metal expand and contract at different rates in response to changes in temperature, which can loosen the joint. Secondly, the metal can draw moisture condensation into the joint, perhaps causing decay in the wood or corrosion in the metal itself.

Figure 4-35: The timber framed New Inn, Herefordshire, England. (Adapted from Timber Framed Buildings, *Arts Council of Great Britain.)*

There are many ways to enclose a timber frame, but the most attractive from the standpoint of speed of construction and energy efficiency is provided by foam-core stress-skin panels. These panels are typically composed of urethane or polystyrene foam sandwiched between ½-inch gypsum drywall on the interior and waferboard or plywood on the exterior. They are designed to resist the strains of tension and compression and can be used to span the 4- to 8-foot distances between timbers. A 4-inch-thick panel with a urethane foam core can have an aged R-value as high as 30. Always find out what the "aged" R-value is, since the R-values of some foams degrade over time.

Since these panels are spiked to the outside of the timbers, they provide an uninterrupted foam blanket around the frame. Nowhere does a framing member pass through the

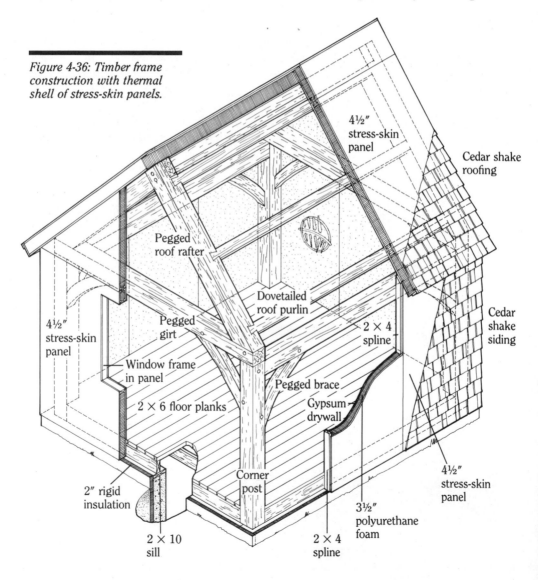

Figure 4-36: Timber frame construction with thermal shell of stress-skin panels.

4½" stress-skin panel

Cedar shake roofing

Pegged roof rafter

Dovetailed roof purlin

Cedar shake siding

4½" stress-skin panel

Pegged girt

2 × 4 spline

Window frame in panel

2 × 6 floor planks

Pegged brace

Gypsum drywall

Corner post

4½" stress-skin panel

2" rigid insulation

2 × 10 sill

3½" polyurethane foam

2 × 4 spline

shell to create a thermal bridge to the outdoors. The connections between the panels are made with wood splines or tongue-and-groove edges in the foam, and openings for doors and windows are reinforced with 2 × 4s that are recessed into the panels. If urethane foam panels are used, no additional vapor barrier is needed. If you anticipate a high-moisture problem or if your panels have polystyrene cores (polystyrene does not act as a vapor barrier), use a good-quality vapor barrier paint. Timber framed homes built with foam-core panels are extremely energy efficient, and they are fully compatible with passive solar heating.

Unless you are an accomplished woodworker, we suggest you hire an expert to cut the frame (you may be able to serve as helper) or buy a precut kit. Timber framers can be found all over the country, and most are highly skilled. Although the joinery is very exacting work and takes some time to do, many timber framers do the cutting in their shops and fit and raise the frame on your foundation, often with a crane, so the frame can be erected and wrapped with panels in just a few weeks. For the name of a timber framer near you, contact the Timber Framers Guild of North America, RR 1, Box 207, Alstead, NH 03602.

Codes and Financing

It's unlikely that you'll run into as much resistance from building officials and lenders over timber frames as you might over domes or earth-sheltered houses. Some officials may hesitate before approving foam-core panels, if they are unacquainted with these items, but the panel manufacturer can supply you with any information you need to satisfy the officials. The frame will probably have to be engineered, but the framer you work with undoubtedly has experience with engineering timber frames.

Timber frame homes are so stunningly beautiful that any concerns your lender might have about marketability should easily be put to rest. These homes, by most estimates, cost about 10 to 15 percent more than comparably finished conventional homes, but to many people the fine craftsmanship makes them worth the cost. The energy efficiency of a timber frame wrapped in insulating panels will also assure good marketability.

Other Methods

The unconventional construction methods we have already listed are generally the most promising for owner-builders. However, a few additional methods are also applicable in some circumstances. We will discuss these methods next.

Pole Houses

If you grew up on or around a farm, you are probably at least passingly familiar with pole barns. Pole houses are built in much the same way. Pressure-treated poles are sunk in the ground (usually they rest on gravel or concrete pads), or they are bolted to concrete piers that rise above the ground. The supports for the floor framing are bolted to the poles. Walls are designed to connect the poles, thus providing bracing, or they can be allowed to "wander" independently, providing no structural reinforcement for the poles.

The advantages of pole houses include relatively low construction costs, adaptability to difficult sites, design flexibility, readily available materials, and superior resilience in earthquake-prone areas. Since the poles form the foundation as well as the structure, the cost of building a complete perimeter foundation is eliminated. An added benefit is that extensive excavation is unnecessary, which means there is minimal disruption to the site. Pole homes are particularly well suited to very steep sites and ocean or lakefront properties in mild climates. They are also appropriate for hot, humid climates where the home can be cooled by outside air introduced from below the living spaces.

You can expect the same resistance from lenders and building officials as with any other unconventional building technique, so do your homework to avoid delays and other frustrations. If you plan to build in a rural area where barns and other utility buildings are built with poles, local officials may already be familiar with the method, even if they're not used to seeing homes built with it.

Cordwood Masonry

Also known as log-end, stackwood, cordwood, or stackwall construction, cordwood masonry is one of the least-known building alternatives. The method is just what the name implies—pieces of seasoned firewood are laid up in mortar to form the walls of a building. In effect, you create a carefully structured woodpile, filling the spaces between the logs with sawdust and mortar on both sides of the pile.

Since in many areas wood is a plentiful and renewable resource, and cutting and splitting firewood are skills within reach of any able-bodied person, a cordwood home may be the ideal answer for those with more time than money. The prospect of building a home of indigenous materials that is energy efficient, attractive (assuming a rustic aesthetic appeals to you), and low-cost is very appealing. Schools offering instruction in cordwood construction are listed in Appendix B.

Cordwood construction is likely to raise more eyebrows at the building department and lending institution than the other alternatives we've mentioned. Since these homes can be very inexpensive to build, you may be able to eliminate the potential hassles with the lender by paying as you go. The building department will want proof that the building will satisfy local building and energy codes.

Kit Homes

The proliferation of kit-home manufacturers is certainly a positive indication of the popularity of the owner-builder concept. Housing consumers can now choose from a staggering array of kits for everything from log homes to timber frame or pole homes. The attraction for the novice builder is obvious. Most kit-home manufacturers offer plans, many will customize the plans to meet the needs of the customer, and some will even assemble a materials list from the client's design. The kit-home dealer then figures the materials for the house, gives the client a price on those materials, and delivers the home as a package of materials that the buyer assembles or hires someone to assemble.

Kit homes certainly offer advantages. Materials are generally high quality, and if your dream house includes materials that aren't readily available in your area, a kit may be the answer. No running back and forth to the local building material yard, no chasing down bids from different suppliers, and no need to make up your own materials lists. If the kit is well designed and planned, you may be able to put it together rapidly with minimal building skills, and some manufacturers even include tools in the package.

Financing is a major marketing tool for some kit-home manufacturers. As we'll discuss in chapter 6, construction financing is one of the most difficult obstacles for owner-builders, so a kit-home builder with construction financing for customers at competitive rates is sure to be popular. Some companies offer plans that require no down payment or only a minimal down payment. For some people, this may be one of the few ways they can afford a home of their own.

Unfortunately, some kit-home companies treat their customers less well than others do. Clearly, in buying something as important—and complex—as an unassembled house, you will want to take every precaution to make sure you're treated well. Here are some things you'll want to check out before you commit to the purchase of a house kit.

1. Make sure you know *exactly* what you're getting for your money. Is it just the shell or are finish materials also included? How about mechanical systems? Does the manufacturer offer the options you want in your home? Is there any flexibility in what you can buy from other sources if the manufacturer doesn't have what you want? Sit down and list everything you will need that is not included in the kit and figure the cost. It might also be instructive to take the final materials list to suppliers and get competitive bids to see how much of a premium you're paying for the kit.

2. If you are uncomfortable with any part of the standard agreement the manufacturer uses, or if you have additional concerns you feel should be addressed, write an addendum designed to become part of the contract. Many contracts are not healthy agreements for consumers, and too many consumers don't notice until they run into trouble. Your expectations of the performance of the manufacturer and the limits of that performance should be clear to both you and the dealer. If you can't get the kind of information you need from the salesperson or dealer, or if you feel you are getting vague or evasive answers to your questions, find another kit dealer. Get a guarantee in writing that the manufacturer will replace any defective parts of the kit.

3. If the kit-home manufacturer advertises itself as supplying high-quality materials, make sure the quality is specified in your agreement and that the kit the company delivers to you contains materials of the specified quality.

4. Don't forget the non-house costs. Roads, water supply, septic or sewer, site preparation, survey costs, permit fees, sidewalks, driveways, etc., are probably not included in the cost of your kit, and therefore are not included in the amount of the construction loan the kit-home dealer is making to you. The money for these extras—and it can add up to many thousands of dollars—is going to have to come out of your pocket, so plan for it. The responsibility for making sure all these details are taken care of is yours, although an experienced dealer can offer valuable assistance.

5. Get a list of local owner-builders that have worked with this dealer. Talk to these individuals to get a feel for what the company is like to work with. Ask whether they feel their expectations were met, whether they received competent technical support during the building process, and ask them how problems were handled. If there is an owner-builder organization in your area, call them to see if they or any of their graduates have had any ex-

perience with this company. If a dealer doesn't want to give you the names of former customers, find another dealer.

6. Pin the dealer down regarding how much support you'll get during the building process. Can you call with questions that come up on a day-to-day basis? If the company provides a building manual, read it carefully to get a feel for how complete it is.

7. Don't use the fact that you're purchasing a kit as an excuse for skimping on planning the project. Take the time to be certain that this is really the house you want. Take classes, attend lectures and trade shows, and read everything you can find on the housebuilding process. Some kit-home manufacturers give classes and workshops on assembling their product, but the quality varies widely. Take a trip to the local building department with your plans and specifications to assure that the house meets local codes. If you plan to hire out any of the work, check with the tradespeople to make sure they're comfortable with the materials offered in the kit.

CHAPTER 5

Selecting Materials

There are many products and materials available that are uniquely suitable for owner-builders. In this chapter, we will describe some materials that we think you should take a look at. We encourage you to read through the entire chapter even if you are depending on subcontractors to provide materials. You may find items in these pages that your subs are unaware of but that suit your situation perfectly.

FOUNDATIONS

To some degree, the success of your entire project depends on the integrity of your foundation. The foundation system you end up with will be determined by a number of factors. If there is expansive or weak soil or a high water table on your lot, the foundation will have to be engineered and is likely to be more expensive than a standard spread footing and stem wall (see figure 5-2).

Expansive soils (clays that expand when they get wet) or weak soils may require pouring the foundation on caissons and void forms rather than a spread footing. A caisson

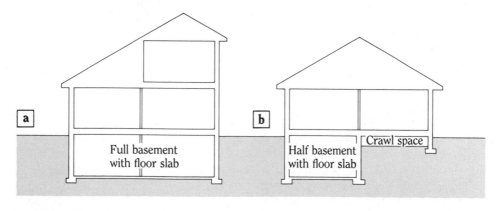

Figure 5-1: Cross sections through houses with a full basement (a) and a half basement (b).

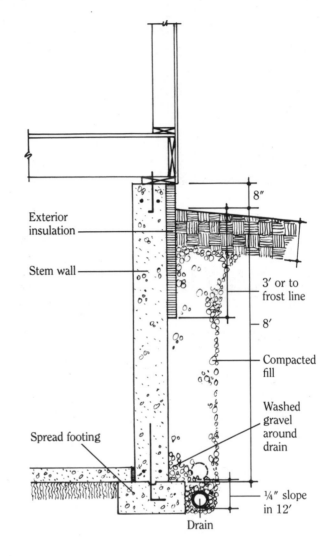

Figure 5-2: Cross section of a typical full basement with a poured concrete foundation wall and spread footing. Moisture protection is provided by a drain tile and impervious topping on backfill.

Exterior insulation

Stem wall

Spread footing

8"

3' or to frost line

8'

Compacted fill

Washed gravel around drain

¼" slope in 12'

Drain

is a concrete pier made by drilling a hole in the ground, setting a reinforcing bar in place, and pouring concrete. The holes are drilled to sufficient depth to assure that the caissons rest on stable soil. Void forms, made of polyethylene-wrapped corrugated cardboard, are placed in the bottom of the concrete forms for the foundation walls. The void forms go between the caissons before the walls are poured. When the concrete forms are removed, the plastic liner is slit, and the cardboard eventually disintegrates, leaving a void to allow for soil expansion without disturbing the wall.

Poured Concrete

Concrete is the standard foundation material. It is made of fine aggregates (sand or rock screenings), coarse aggregates (crushed stone), Portland cement, and water. The aggregates are proportioned so that the finer ones fill the gaps between the larger ones. The aggregates also serve to make the concrete more economical, since they are considerably less costly than cement, and they reduce the amount of shrinking and cracking that occurs as the mixture cures. Portland cement is made by combining a number of minerals, including limestone, iron ore, sand, etc., firing them in a kiln, and pulverizing the resulting "clinkers." The American Society for Testing Materials (ASTM) recognizes five types of Portland cement, of which Type I and Type III are the most common. Type I is what you're likely to use in residential applications. Type III is designed to cure quickly and is useful when the forms must be removed early, and/or the concrete must assume its full load soon after the pour.

Concrete does not dry out—it "cures." Water combines with the cement in a chemical reaction called hydration that binds all the aggregates together. About half the water in the mix is permanently incorporated into the concrete. The water you use should contain no oil, alkali, or acid, and drinkable water is best. The amount of water affects the workability of the mix. Wetter concrete may be easier to pour and handle, but adding too much water to the mix can significantly weaken the finished product.

The relative sloppiness of the mix is specified as "slump." If your foundation plans have to be engineered, the engineer may require a slump test before certifying the structure as sound, although this is uncommon for a residential building. A slump test is done by filling a truncated cone (12 inches high with an 8-inch base) with concrete, inverting the cone, and placing it beside the resulting pile of concrete. The distance the concrete sags

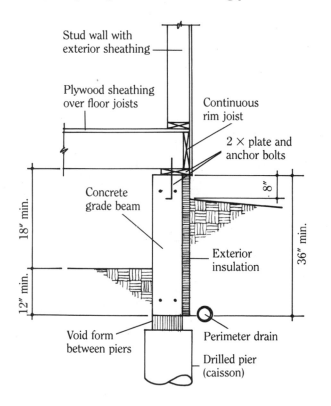

Stud wall with exterior sheathing

Plywood sheathing over floor joists

Continuous rim joist

2 × plate and anchor bolts

8"

Concrete grade beam

18" min.

Exterior insulation

36" min.

12" min.

Void form between piers

Perimeter drain

Drilled pier (caisson)

Figure 5-3: Cross section of a typical crawl-space wall made of poured concrete and supported by caissons and void forms. (Redrawn with permission from E. John Stuart.)

from the top of the cone, measured to the ¼ inch, is the slump. The greater the slump, the sloppier, and weaker, the mix. For residential footings, walls, and slabs, a 4- to 6-inch slump is usually acceptable.

Concrete can be ordered with a variety of admixtures that will cause it to cure more slowly or more quickly, help prevent freezing, and reduce the amount of water needed. It is also possible to mix in dyes to produce colors other than the familiar gray. Many people dye their concrete slabs and either stamp or cut patterns into the surface to resemble tile or stone. The effect can be quite impressive and is considerably cheaper than installing tile or stone floors separately.

Because cement contains lime, a strong alkali, it can be hard on skin. You should wear rubber boots and gloves when pouring concrete. Washing your hands with vinegar will neutralize any lime that reaches your skin.

While the skills for pouring footings and stem walls are probably well within reach of most motivated novices, pouring and finishing a slab is another story. Get a professional to work with you for at least the day of the pour, since it's not uncommon for a slab to cure more quickly than you can finish the pour, especially in hot, dry weather. Once water is added to the mix, the curing process begins. If the water is added at the concrete company

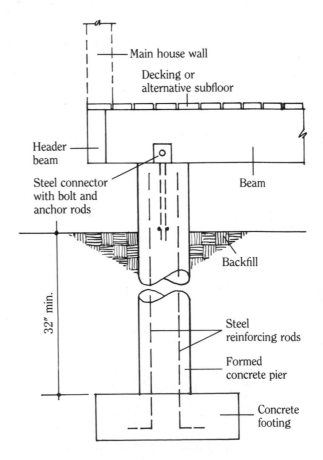

Figure 5-4: Cross section of a concrete pier foundation on spread footing.

and the truck must travel a long distance to your site, the mixture may be "hot," or starting to harden even before you begin to pour it. Some people add water at this point to make the concrete more workable, but adding too much water can weaken the slab. There are ready-mix companies that use trucks to carry aggregates, cement, and water in bays in the truck and then mix the batch at your site.

After it is poured, concrete should cure slowly to assure a satisfactory appearance and optimum strength. Exposed surfaces should be kept moist for at least three days after the pour. If a surface is allowed to cure quickly, it will shrink, resulting in a dusty surface and hairline cracks. A fresh pour must be protected from rain and free water, however, since they can have the same effect as adding too much water to the mix. The concrete must also be kept from freezing, since uncured concrete that freezes will be weak and is likely to crack and develop scales on its surface. Because concrete generates heat as it cures, this is often simply a matter of covering the pour to keep this heat in.

Concrete Block

Concrete blocks are another option for foundation material. They are available in enough sizes, shapes, colors, textures, and profiles to satisfy the most creative designer. They eliminate the need for complex formwork, so one person alone can easily build a wall.

In the United States, concrete blocks are manufactured to conform to the requirements of ASTM, which grades the units according to the intended use and the degree of moisture control desired. Grade "N" is for general use in exterior walls above and below ground level where the wall will be exposed to moisture. Grade "S" is limited to use aboveground in exterior walls with weather-resistant coatings and in walls that will not be exposed to the weather.

Mortar-Jointed Block

The conventional way to build with concrete blocks is to set the blocks using cement mortar, leveling each course carefully. Block foundation walls are laid on a poured concrete footing.

If you want a mortar-jointed concrete block foundation, we recommend that you hire a mason to build it for you. Unless you have some experience in laying block, you probably won't become proficient until the job is almost done. A journeyman mason can average 200 blocks a day, and you'll be doing well to average half that amount even after a couple of days of practice. If, for your own reasons, you want to build a mortar-jointed block foundation yourself, consider either working with a mason to learn the proper techniques or hiring a mason for a few days to get you started.

Surface-Bonded Block

Another method of building foundation walls with concrete block is to dry-stack the blocks (using no mortar) and then surface-bond them. The blocks are stacked in a "running bond" (every block overlaps two blocks below it), then the surface-bonding mix is

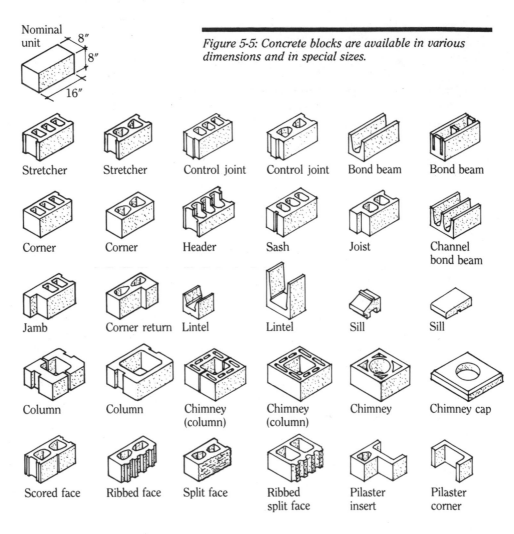

Nominal unit 8" 8" 16"

Figure 5-5: Concrete blocks are available in various dimensions and in special sizes.

Stretcher	Stretcher	Control joint	Control joint	Bond beam	Bond beam
Corner	Corner	Header	Sash	Joist	Channel bond beam
Jamb	Corner return	Lintel	Lintel	Sill	Sill
Column	Column	Chimney (column)	Chimney (column)	Chimney	Chimney cap
Scored face	Ribbed face	Split face	Ribbed split face	Pilaster insert	Pilaster corner

troweled onto both sides of the wall. The bonding mix is a concoction of Portland cement, lime, calcium chloride (for fast setup and a harder finished product), calcium stearate (to make it more waterproof), glass-fiber filament chopped into ½-inch lengths, and water. The glass fibers bridge the cracks between the blocks so the finished wall will have about six times the strength of a mortar-jointed block wall. The first course is set in mortar on a concrete footing and carefully leveled, and subsequent courses are leveled with metal shims as they are laid up. The hollow cores of the blocks are filled with concrete, following the specifications of the structural engineer or building department. Many building departments allow the use of recycled blocks for surface-bonded walls, which can result in significant savings.

Compared to mortar-jointed block walls, surface-bonded walls are faster and cheaper to build, stronger, more watertight, more attractive. The bonding mix can be purchased as a dry premix with color added.

Should you decide to build surface-bonded walls, keep in mind that a standard "8-inch by 16-inch" concrete block is really 7⅝ inches by 15⅝ inches, to allow ⅜ inch for the mortar joint. Be sure to take this into consideration when you figure your wall height and length. Also, conventional concrete blocks are not made to very precise measurements, since it is assumed that they will be laid up in mortar, which evens out irregularities. It may take a considerable amount of shimming to level the courses. Discrepancies over ⅛ inch should be shimmed with mortar.

Interlocking Block Systems

An even more attractive option for novice builders is one of the new interlocking block systems. These blocks are also stacked without mortar, but the big advantage they offer is that they are milled to much closer tolerances than conventional concrete blocks, so shimming is less necessary. They are a true 8 inches by 16 inches, so the height of the wall is easier to estimate.

Some of the manufacturers of dry-stack blocks recommend that their product be surface-bonded; others cast tongues and grooves into the blocks to keep the stacked wall rigid; and still others use plastic splines or rings to keep the blocks aligned. One system we have been impressed with is MCIBS, an acronym for Mortarless Concrete Interlocking Block System. Introduced in 1980 by MCIBS, Inc. (130 South Bemiston, St. Louis, MO 63105), these blocks feature tongues and grooves and are manufactured to tolerances of 0.03 inches. The ease with which these blocks can be used makes them ideal for unskilled workers. They are at least four times faster than mortared block to build with. They are considerably more expensive than conventional concrete blocks, but if you figure in labor costs and all materials, a wall built with interlocking blocks can actually end up being 10 to 15 percent less expensive than a mortared block wall.

All-Weather Wood Foundations

It might sound ridiculous to build a house's foundation out of wood, but all-weather wood foundations (AWWF) have been extensively tested for nearly 20 years and are now supporting about 50,000 buildings around the country. The system is recognized by all the model building codes, the federal agencies regulating housing, major lending and mortgage insurance institutions, and warranty and fire insurance institutions. Pressure-treated lumber and plywood (wood that has been impregnated with a wood preservative) are used for the foundation walls, which stand on gravel footings. The preservative in the wood makes it resist decay and insects.

Dealers claim an AWWF costs only two-thirds as much as a concrete foundation. Assembling the walls requires only basic carpentry skills, good news for most owner-builders. Wood foundations also adapt more easily to odd or irregular shapes than poured concrete or block walls, and they are insulated and finished like any other wood-frame wall. Wood foundations can be built in any weather, although the installation will certainly go more smoothly and efficiently if the temperature is above 40°F and it isn't raining.

The walls can be prefabricated to save time at the job and to minimize the amount of cutting you need to do. Ideally, you should not cut any of the lumber in an AWWF. The biggest disadvantage to an AWWF is that the wood is preserved with copper arsenates. There is some evidence that these salts cause serious health problems. During installation, wear goggles, a dust mask, rubber or vinyl-coated gloves, and heavy coveralls. Never saw, sand, or plane pressure-treated wood indoors, as the arsenic-laden sawdust will get into the air. We also encourage you to cover treated wood inside the home to assure that occupants won't come into direct contact with it. Don't burn the scraps, as the smoke will contain arsenic particles. Scraps must be buried.

To be effective, all surfaces of an AWWF must remain treated with preservative. If it is necessary to cut or drill any of the lumber, the affected pieces should be brushed, dipped, or soaked until the wood absorbs no more preservative. Take precautions while doing so. Or better yet, as we advised above, avoid cutting into pressure-treated lumber in the first place.

An AWWF must have an effective drainage system. The wood foundation and the concrete slab that (usually) forms the floor of the basement are typically built on a leveled gravel pad, which is an integral part of this drainage system. A 6-mil polyethylene plastic sheet covers the exterior of the below-grade portions of each foundation wall. This plastic film directs water to the gravel footing and drain tile, so that it can then be drained to the surface of the ground several feet away from the house. Downspouts should drain onto splash blocks to direct water away from the building, and the ground should slope away from the house. In basements, a sump should also be installed and drained or pumped to the outdoors.

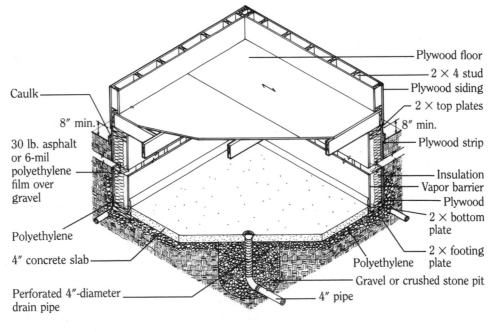

Figure 5-6: Typical all-weather wood foundation. (Adapted with permission from the American Plywood Association.)

Careful backfilling, in which care is taken not to damage the polyethylene, is usually done in 6- to 8-inch layers after the basement floor has been poured and the first floor framing and plywood subfloor are in place. The section of the backfill nearest the footing is filled with the same gravel that makes up the footings, and it is covered with strips of 30-pound asphalt-impregnated roofing felt. These strips allow water to seep through to the drain tile but prevent soil from filtering down to clog the gravel.

Moisture-Proofing and Insulation

Most building codes require that foundation systems be moisture-proofed below grade. Traditionally, asphalt products, which are petroleum based, were used for this purpose. There is a problem, however. Rigid plastic foam insulations are now in wide use as foundation insulation, and the petroleum solvents in asphalt foundation coatings dissolve the foam. If the coatings are allowed to cure before the foam is installed, the problem may be minimized, but there are also other alternatives.

First, there are products that perform as both waterproofing and adhesive, and that are compatible with foam insulations. Second, there are many moisture-proof coatings that are compatible with foams but that don't serve as adhesives, so the insulation must be attached mechanically or with a separate adhesive. Check with your local building materials supplier for the brands available in your area.

Drains

The best way to keep your basement or crawl space dry is to use an integrated approach that takes site conditions into consideration. An important part of this approach is the use of footing drains to keep water from building up next to the foundation wall. The drains should be laid in gravel and then covered with gravel. In turn, the gravel should be covered with building paper or strips of polyethylene to prevent soil from filling the spaces between the stones. The gravel must always be larger than the holes in the drain tile, and, if you decide to use PVC tile with holes on one side, install the tile with the holes down.

If your soil drains well, you will need only gravel around the drain tile, but if the soils on your site are expansive or don't drain well, consider backfilling with a granular fill at least halfway to the surface of the ground. In any case, the backfill should be placed in 6-inch-deep layers that are individually compacted to avoid settling later. Settling can drag insulation off the wall, damage the waterproofing on the wall, and alter surface conditions so that the ground slopes toward the house, encouraging water to run down along the foundation wall.

Between the top of the gravel and the soil you place above it, backfill with water-impervious clay. Slope everything away from the house at a minimum drop of 3 inches in 10 feet. Ideally, the drains should be run out to the surface of the ground, but if that isn't possible, a dry well is usually sufficient. Downspouts should never empty directly onto the soil around the foundation. At a minimum, use splash blocks or a small drainage system to take the water from the downspouts away from the foundation.

Insulation

There is some debate about whether it is necessary to use extruded polystyrene as insulation below grade or whether expanded polystyrene (EPS, otherwise known as beadboard) is sufficient. Extruded polystyrene is about twice as expensive as EPS. In our view, there should be no problem using beadboard in a conventional situation with good drainage. In extremely cold climates, "impervious" extruded polystyrene is preferable to prevent frost damage.

There are products now on the market that insulate as well as facilitate drainage. GeoTech Insulated Drainage Board or Panel (GeoTech Systems Corporation, 118 Acacia Lane, Sterling, VA 22170) is a product composed of large, high-quality EPS beads bonded with a specially developed, waterproof asphalt adhesive. The bonding of the beads with the adhesive creates channels throughout the board that allow groundwater to drain down to the drain tile at the footing. Plastic, foil, or some other moisture barrier can be placed on the foundation side of the panels to keep the groundwater from coming into contact with the foundation wall. The earth side of the panels can be covered with a fabric that will keep fine soil from clogging the panels. This arrangement prevents water from building up and exerting pressure on the wall.

The panels protect the foundation wall and its waterproofing from damage during backfill, eliminating the need for the labor and expense of extra protective layers. They are light and easy to handle and install, usually requiring only GeoTech mastic without additional mechanical fasteners. They have a dry R-value of 3.5 per inch.

Another product that both insulates and promotes drainage is a rigid fiberglass basement insulation, called Baseclad (in Canada) and Warm-N-Dri (in the United States). It features thin, discrete planes of fiberglass layered in such a way that if the product is installed vertically, water contacting the surface of the board is channeled down to the footing drainage system. It has good resiliency and compressive strength, making it particularly useful in areas where expansive soils are a problem.

It is also possible to insulate a basement or crawl space from the inside. A stud wall is power-nailed to the inside surface of the concrete wall and then insulation batts are hung between the studs. It is also possible to install rigid insulation between furring strips. If you use rigid insulation, most codes require that you cover the foam with a fire-rated material, since the foam will emit poisonous fumes if it burns.

We recommend that you insulate the outside of the foundation for several reasons. First, there is the potential for water damage in a basement that has been insulated from the inside. Moisture may build up between the insulated stud wall and the concrete wall and pool on the floor between the two. Second, when a concrete wall is insulated on the outside, the concrete is inside the heated space of the building, so it is warmer and less prone to damage from freeze/thaw cycles. Also, by virtue of its considerable thermal mass, the concrete will help even out the day-to-night temperature swings in the basement.

In cold areas where the soil retains a great deal of moisture, insulating foundation walls may expose the soil around the foundation to freezing temperatures, which can cause heaving. When basements aren't insulated, the heat lost to the surrounding soil may keep the soil from freezing. If wet soil freezes, it could expand enough to exert lateral pressures on the foundation walls that could crack the walls. If this is a potential problem in your area, you may want to backfill around the walls with gravel or some other granular material that doesn't retain moisture. See the specifications we gave earlier for backfilling around an AWWF.

FRAMING AND SHEATHING

Standard 2 × 4 stud construction is the most common way to build a home in the United States, for several reasons. It is easy to learn and does not call for great skill. Information about it is widely available, and each framing member is easy to handle, so the work goes quickly without the need for heavy equipment. The lumber is widely available and relatively inexpensive, and both exterior and interior finish materials are manufactured to fit this framing style.

When shopping for framing lumber, look for wood that is as dry as you can find. Drier wood is more dimensionally stable, lighter, and easier to work with. In today's tighter homes, movement from shrinkage can open carefully sealed joints, particularly around doors and windows. You may not have a choice, however. Much of the available framing lumber these days is stamped S-grn, or "surfaced green," which means its moisture content is above 19 percent. Remember, too, that different wood species have different strengths, and it's a good idea to design for the weakest framing lumber available in your area to assure that the lumber you get can handle the loads imposed on it.

Insulation

Energy-efficient building practices require adequate levels of insulation. Most people are now familiar with the concept of "R-value," or the resistance of a material to heat transmission. Products manufactured as insulating materials have exceptionally high R-values.

Whether or not an insulation achieves its rated R-value depends on several factors. Motionless air is one of the best insulators around at about R-5.5 per inch. With that in mind, you might think that an empty 2 × 4 stud space should have an R-value of 19.25. The problem is that the air in the space is not really motionless—its effective R-value is actually only about 1. So to prevent heat from escaping through your walls, you will need to install insulation. We will list your options below.

Fiberglass and Rock Wool

The advantages of fiberglass include wide availability, low cost, and compatibility with conventionally framed structures. Batts and blankets are made to fit in standard 16- and 24-inch framing cavities, so installation is easy. (Cutting the batts or blankets to fit irregular spaces can be a chore, however.) They have an R-value of 3.2 per inch.

"Loose-fill" fiberglass can be blown or poured into cavities and attics. Care must be taken in vented attics so that the wind doesn't blow the fiberglass around, leaving some areas devoid of insulation. An adhesive binder can be added to the loose-fill to keep it in place once it has been blown. Loose-fill fiberglass rates approximately R-3 per inch, depending on how and where it is installed. Like other loose-fills, it can be a boon in irregular spaces, eliminating the need for cutting of insulation batts, blankets, or boards.

Rock wool was the first insulating material manufactured on a large scale, but its use has declined over the years. It has many of the same characteristics as fiberglass and is available in batts and blankets (R-3.4 per inch) and loose-fill (approximately R-2.9). Both fiberglass and rock wool are noncombustible, although the binder in fiberglass can develop toxic fumes when burned.

Cellulose

Cellulose is paper or virgin wood that has been shredded and milled to produce a fluffy, low-density insulating material. Chemicals are added so that the insulation resists fire, water absorption, and fungal growth. Slightly moistened cellulose is blown into wall cavities where it assumes a papier-mâché-like consistency. It has a higher R-value per inch than loose-fill fiberglass and rock wool (approximately R-3.5); is more fire resistant because of its greater density; fills all the little nooks and crannies that batts or blankets sometimes miss; and resists the movement of air through the wall better than batts and blankets do.

Add to this the fact that cellulose insulation is usually a recycled product, that fiberglass takes 7 to 10 times as much energy to produce, and that superior levels of sound resistance are possible with cellulose, and you can see that cellulose is a very attractive option. Concerns about the fire resistance of this product have been laid to rest by diligent quality control and independent testing, and the cost is now generally competitive with fiberglass. A potential disadvantage for owner-builders, however, is that for best results, cellulose should be installed by a professional.

Foams

The most common foam insulations are polystyrene, polyurethane, and polyisocyanurate. All are available as boards sized to fit between standard framing members, and polyurethane and polyisocyanurate can also be blown in place. Cutting insulation boards is easier than cutting batts or blankets, but cutting a large number of them can be taxing: You must make the cuts precise to avoid leaving gaps.

Polystyrene is available as expanded polystyrene (EPS or beadboard) or extruded polystyrene. The extruded variety has a higher R-value per inch (R-5), because it contains a mixture of air and fluorocarbons (which have a higher R-value than air), while EPS (R-4) contains only air in its cells. Extruded polystyrene is also stronger and about twice as expensive.

These foams are often used as sheathing on the exterior of frame walls, as insulation on the exterior of masonry walls, and as foundation insulation. EPS is permeable to water vapor, so using it as exterior sheathing does not provide a vapor barrier. The other foam boards do act as vapor barriers, which may be a disadvantage if you live in an extremely cold or exceptionally humid climate.

Polyurethane and polyisocyanurate (R-6 and R-7.4, respectively) have the highest R-values per inch of the commonly available foam insulations. They can be used as exterior

Table 5-1 COMMON INSULATIONS

Type	Form	R-Value*	Cost	Combustible?
Fiberglass	Blanket/Batt	3.2	Low	No, except facing
	Loose-fill	3	Low	No
Rock wool	Blanket/Batt	3.4	Low	No, except facing
	Loose-fill	2.9	Medium	No
Cellulose	Loose-fill	3.5	Low	Yes, unless treated
Extruded polystyrene	Board	5	High	Yes
Expanded polystyrene	Board	4	Medium	Yes
Polyurethane	Board	6	High	Yes
Polysio-cyanurate	Board	7.4	High	Yes

Per inch thickness.

sheathing or they can be installed on the interior surfaces of walls. In situations where high humidity is a problem, interior installation may be preferable, since these foams block the passage of moisture and form a vapor barrier. Bear in mind, however, that these foams release highly toxic fumes when they burn. Most building codes require that the foams be covered with gypsum board or another fire-retardant material when the foams are installed on the interior surfaces of walls.

Air/Vapor Barriers

No matter how much insulation you pile into your ceiling, walls, and floors, the house will still waste energy if air that you've paid to heat is allowed to leak out of the house. A related issue is that moisture penetrating into the shell of the house can damage the structure and lower the R-value of the insulation.

Water vapor enters a wall either by diffusion or by convection. Water molecules are smaller than molecules in the air, and they can move through materials that block airflow. This passage of water molecules directly through a material is called diffusion, and it occurs when the relative humidity on one side of a material is higher than on the other side. The water vapor moves from the area of higher concentration, usually the warm inside of the house, to the area of lesser concentration: the wall cavity. Installing a vapor barrier on or near the inside surface of the wall will control this sort of diffusion.

Diffusion is only part of the problem, however. Most moisture that enters a wall cavity is carried there by warm air moving through cracks, holes, and seams in the wall. This air movement, called convection, can be controlled by making the vapor barrier as airtight as possible—hence the term air/vapor barrier. Traditionally, builders used polyethylene sheets (6 mil or greater) for air/vapor barriers, but more and more builders are using specialized products that should last longer and work better. A few of these products are Tu-Tuf Moisture Vapor Barriers (Sto-Cote Products, Inc., P.O. Box 310, Richmond, IL 60071), Rufco 300 and 400 (Raven Industries, Inc., Box 1007, Sioux Falls, SD 57117), and Super-Sampson (Poly Plastic and Design Corp., 1920 East Pleasant Street, Springfield, OH 45501). These are cross-laminated, high-density polyethylene sheets, which are much stronger and more durable than conventional polyethylene. If you can't find these products in your area, write the manufacturer or try a greenhouse supplier.

Air/vapor barriers should be installed on the warm side of the insulation. In most climates in the United States, this means installing the barrier on the inside of the wall. If you live in a hot, humid climate, however, you may want to install an air/vapor barrier on the exterior of the wall, to keep outdoor humidity from entering the wall. Check with experts in your area.

ROOFING

A number of factors will affect your choice of roofing materials, among them cost, weight, fire resistance, durability, and the pitch of the roof. Here are a few tips:

■ It is useful to think in terms of life-cycle costs as well as purchase costs, since roofing is a fairly major home maintenance task. So when comparing roofing materials, think of how much you will have to spend to maintain and ultimately to replace the roof you select. Some metal, concrete, tile, and slate roofs will last indefinitely, assuming the fasteners don't rust or rot away and the roof isn't physically damaged.

■ Most roofing materials aren't heavy enough to pose a threat to your roof framing. But concrete, tile, and slate roofs are so heavy that you'll need to plan extra-sturdy framing for them.

■ Underwriters' Laboratories tests roofing materials for fire resistance and assigns class ratings to them. Class A is the highest rating, for fairly fire-resistant materials, and Class C is the lowest. Some materials, such as untreated wood shingles, have no rating at all, since they burn readily.

■ Some roof materials may be more durable than others in your particular situation. If it is very windy in your area, for instance, you'll need a roof that can withstand high winds.

■ The pitch of your roof will also be a factor in determining what materials you can use and how well the roof will shed snow and water. In discussing specific materials, below, we give the recommended minimum slope for each material, but many can be used on a shallower slope if alternative application methods are followed.

Asphalt Shingles

Asphalt shingles weigh 200 to 300 pounds per "square" (a square is the amount of roofing material needed to cover 100 square feet). They are available with an organic felt base or a fiberglass base. The newer fiberglass-based shingles are lighter, stronger, more fire resistant (Class A, compared to felt shingles' Class C), and they last a little longer (about 25 years to felt's 20).

Both felt-based and fiberglass-based shingles are easy to install, and they come in a variety of colors. (Bear in mind that lighter colors stay cooler in the summer and help the roof last longer.) Most building departments require at least a 4-in-12 pitch (the roof rises 4 inches for every 12 horizontal inches) for asphalt shingles. Asphalt shingles that have the uneven appearance of wood shakes are also available.

Wood Shingles and Shakes

Wood shingles and shakes are a popular roofing material because they are easy to apply and make an appealing roof, raising the value of the home. The big drawback to wood is its fire hazard. Some shingles and shakes have achieved a Class C rating after treatment with fire-retardant chemicals, but most are unrated—they are highly flammable.

Cedar, cypress, pine, and redwood will all work for roof shingles, but cedar is by far the most popular. Cedar roofs start out a reddish brown color and weather to a silvery gray or light tan, depending on the climate. Always use No. 1 grade cedar. Your building department will require that your roof have at least a 4-in-12 pitch, unless you use a double layer of roofing felt. A wood roof usually weighs between 200 and 300 pounds per square and will last about twice as long as an asphalt roof, all other things being equal.

Concrete and Clay Tiles

Concrete and clay tiles are among the heaviest materials you can use on a roof, weighing anywhere from 700 pounds to 1,600 pounds per square. They are available in a variety of styles and colors, from shingles resembling slate to traditional barrel tiles. They will last indefinitely if installed properly. They require at least a 3-in-12 pitch and may not be a do-it-yourself project. Some of the concrete-tile manufacturers provide wonderfully clear instructions for installing their products, but the advantage to hiring professionals is that they warrant their work, so you'll have recourse if your tiles blow away in the first major windstorm.

Although they can be expensive, perhaps three or four times the cost of asphalt shingles, they create a beautiful, permanent, very fire-resistant roof. Colors are permanent in clay tiles, but concrete tiles tend to fade. Costs vary from area to area and will be lower if you have a manufacturer in your area—the substantial weight of the tiles makes it expensive to ship them any distance.

Slate

Roofing your house with slate makes sense economically only if you live in an area where the material is mined or you have access to used slate shingles. Costs vary dramatically from one area to another, but slate tends to be one of the pricier roofing materials.

Slate roofs are beautiful, however, and it's not uncommon for them to keep a house dry for a century or more. They are fire resistant and require special tools to install, although the skills involved aren't beyond the average owner-builder.

Shipping this heavy material is expensive (slate weighs at least 750 pounds per square). The color of slate shingles can range from grayish blue to red or green. Some slates will change color as they weather.

Metal

Metal roofing is generally stronger, lighter, and more durable than more common roofing materials. It goes up in large sheets with few fasteners, so leaks are less likely to occur and are easier to find if they do occur. Metal shingles and shakes are also available. In areas where fire hazards are high but a shingled appearance is desired, metal shingles offer a lightweight alternative to concrete or tile roofs. Metals used for roofing, from most expensive to least expensive, include zinc alloy, copper, terne-coated stainless steel, stainless steel, terne metal, aluminum, Cor-Ten, and galvanized steel. All of these metals corrode, and they will expand and contract in response to changes in temperatures. Each has its own characteristics, which must be accommodated in the design and installation of the roof.

Zinc roofing weathers to a gray color and can be expected to last more than 100 years under most circumstances. It also tends to sag and creep on the roof, however, which has led at least one manufacturer (W. P. Hickman Company) to alloy the metal with copper and titanium. Zinc is prone to corrosion by galvanic action, so it must be protected from contact with most other metals. Oak, redwood, and cedar contain natural acids that will damage zinc. However, assuming precautions are taken, zinc will form an attractive layer of corrosion that protects the roof from further corrosion, and runoff from the roof will not stain siding or discolor landscaping.

Most people are familiar with the brown-green patina that weathered copper takes on. This patina is a layer of sulphates that protects the copper from further corrosion. Copper roofs can last hundreds of years but will corrode other metals and stain siding and trim. Like zinc alloys, copper is an expensive roofing material.

Terne metal is an alloy of lead and tin on a steel base. When the alloy is placed over stainless steel, no surface finish is required—the terne will weather to a dark gray. Terne over stainless steel is a permanent roofing material, but it is nearly as expensive as copper. Conventional terne metal is considerably less expensive (about half the cost), but it must be finished to prevent rusting, and it needs refinishing every ten years or so.

Stainless steel contains chromium and nickel, which makes it resistant to corrosion and rusting. Stainless steel doesn't corrode other metals, and it costs about two-thirds the price of copper. Some people find stainless steel roofs too bright, and because the material is so resistant to corrosion, the finish will not dull significantly with time.

Aluminum resists corrosion better than most metals, making it a particularly attractive choice along seacoasts where the salt in the air will speed up corrosion. Aluminum will, however, react galvanically with most other metals, especially in salt air, and should be protected from physical contact with other metals. Aluminum is lightweight, but it expands and contracts more than most metals, so installation must be designed with that in mind.

Cor-Ten is a weathering steel manufactured by U.S. Steel. It is designed to rust readily, building up a thick surface layer that protects the steel below from further corrosion. It rusts to an attractive reddish color, and runoff must be directed away from sidings and plantings because it will stain them. Cor-Ten is heavier than other metal roofings.

Galvanized steel is the least-expensive metal roof (about the cost of a premium asphalt shingle roof). Quality varies widely, and the material is available with a variety of coatings that prolong its life. It is light, strong, and available in many patterns that have a clean, attractive appearance. Galvanized steel is ordinary steel plated with zinc, but aluminized steel and aluminum/zinc alloy platings (such as Bethlehem Steel's Galvalume) are also available. The aluminized metal roofing is a good choice in coastal areas, since it's less likely to corrode in salt air.

Other Roofing Materials

Other materials are also available for roofs. Mineral-fiber shingles, for example the asbestos-cement shingles manufactured by Supradur and the lightweight perlite shakes made by Cal-Shake, have a Class A fire rating and cost about twice as much as top-quality fiberglass-based asphalt shingles. They will also last twice as long, making them a cost-effective alternative to an asphalt-shingle roof.

Masonite Corporation makes wood-fiber roofing that resembles cedar shakes, weathers to a silver gray, and comes in 1-foot by 4-foot panels for fast installation. Treated with fire-retardant chemicals, it has a Class C fire rating.

If your house design calls for any flat or nearly flat roofs, you have another array of options. The least-expensive material is mineral-surfaced rolled roofing, which can be used at pitches as low as 1-in-12. Rolled roofing can also be used on pitched roofs where budgets are tight. Some owner-builders apply rolled roofing as a temporary measure to complete their homes on tight budgets, then reroof with higher-quality materials as money becomes available.

Onduline is a corrugated asphalt sheet that has been in wide use around the world for about 20 years. The large size of the sheets (46 inches by 79 inches), variety of colors, modest cost (about the same as premium-quality asphalt shingles), and light weight of the material all contribute to its growing popularity. Corrugated asphalt sheets are effective down to a 1-in-12 pitch.

Synthetic rubber membranes are another choice for flat roofs, and although they are expensive, they are a permanent covering. The seams are glued with a cement that should last at least 20 years, after which time the seams may need regluing. This material is also used as flashing in situations where metal flashing doesn't have the necessary flexibility.

There are flexible roofing materials that are applied as a liquid. They bond to the substrate or underlayment and form a watertight roof. Some require specialized equipment to apply or are franchised only to authorized installers, but others can be applied with a paint roller or other readily available applicator.

SIDING

A number of factors will affect your choice of siding. Appearance, durability, maintenance, and availability of materials must all be considered. The building method you are using will also be a factor. If, for example, your home is rammed earth, a stucco finish makes more sense than furring all the exterior walls with wood and using cedar siding. Other factors you should think about include purchase cost, ease of installation, and combustibility. Occasionally, covenants or architectural guidelines will dictate what siding you can use.

Wood Board Siding

Wood is a popular siding for a number of reasons. It is readily available in most areas, makes an attractive exterior finish, and can be installed relatively easily. Some species—notably cypress, cedar, and redwood—have exceptional natural resistance to insects and rot. Board sidings generally carry a grade stamp that identifies the species and quality of the product.

Moisture content is of special concern when choosing sidings, since unseasoned wood can shrink, cup, and split after it's installed. Wood that is certified kiln-dried is typically the driest—and the most expensive—but it may not be available in your area. Often boards are air-dried to specific moisture contents. For instance, "MC-15" indicates that the board has been seasoned to a 15 percent maximum moisture content, "S-dry" means the wood contains a maximum of 19 percent moisture, and "S-grn" indicates the wood is green or unseasoned (moisture content above 19 percent).

Cedar and redwood are common choices for sidings. They are beautiful, available in a wide range of grades and patterns, easy to work, dimensionally stable when properly seasoned, and they readily accept finishes. Costs vary according to the amount of milling required, how well the wood has been seasoned, and the grade.

The most expensive grades are marked "clear, vertical grain, heartwood." "Clear" indicates the wood is free of knots and other irregularities. "Vertical grain" tells you that the grain runs parallel to the long edges of the board and that this board has the best possible surface for painting. "Heartwood" is the most decay-resistant part of the tree and can be identified by its darker color.

Wood board siding consists of boards installed on the exterior surfaces of a home's walls. Standard patterns include the following: beveled, shiplap, tongue and groove, and board and batten. Most patterns are available in various widths. Generally, boards 6 inches or smaller (nominal size) require one nail at each framing member, while 8-inch or larger boards require two. Nails penetrate one board and the framing, not the board underneath. This leaves the siding a little room to expand and contract with temperature and moisture changes.

Many people prefer the texture and knotty appearance of the less-expensive rustic sidings. These are available in a number of patterns. When you shop for rustic siding, avoid boards that have large, loose knots, large splits, or large bows or twists. Reputable lumberyards will usually take back wood that is unusable. Rustic siding is available seasoned and has a rough (saw-textured) face.

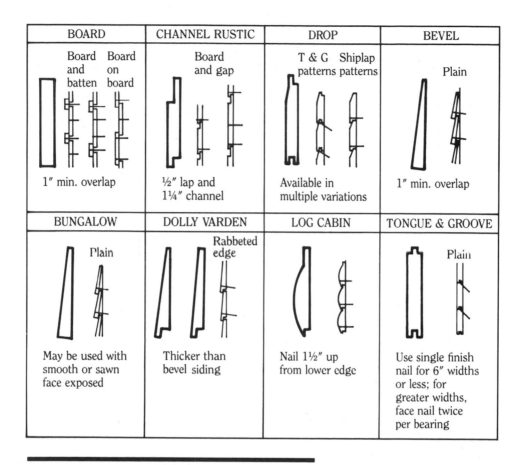

BOARD	CHANNEL RUSTIC	DROP	BEVEL
Board and batten — Board on board	Board and gap	T & G patterns — Shiplap patterns	Plain
1″ min. overlap	½″ lap and 1¼″ channel	Available in multiple variations	1″ min. overlap
BUNGALOW	**DOLLY VARDEN**	**LOG CABIN**	**TONGUE & GROOVE**
Plain	Rabbeted edge		Plain
May be used with smooth or sawn face exposed	Thicker than bevel siding	Nail 1½″ up from lower edge	Use single finish nail for 6″ widths or less; for greater widths, face nail twice per bearing

Figure 5-7: Various wood sidings and their nailing patterns. (Adapted with permission from Western Wood Products Association.)

Cedar Shingles and Shakes

Cedar shingles and shakes also make for handsome, weather-tight siding. They are particularly attractive to owner-builders who are doing their own work because one person can easily install the material. Shakes or shingles are either "single-coursed" or "double-coursed." Double-coursing involves installing two layers or courses of shingles—an undercourse of lower grade (No. 3 or undercoursing grade) shingles and a surface course of higher grade (usually No. 1). This technique makes it possible to leave a larger area of the surface course exposed than would be possible with only one layer of shingles.

Top-quality, noncorrosive nails should be used. Recommendations for using building paper under the shingles vary, so check with local builders or building officials. Shakes and shingles bonded to plywood sheathing are available from Shakertown Siding and Roofing (P.O. Box 400, Winlock, WA 98596) in 8-foot lengths, are self-aligning, and greatly reduce

SINGLE-COURSING

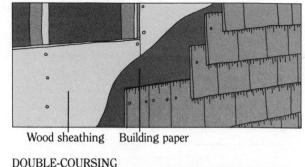

Figure 5-8: Single-coursing and double-coursing of shingles and shakes. (Adapted with permission from Red Cedar Shingle and Handsplit Shake Bureau.)

Wood sheathing   Building paper

DOUBLE-COURSING

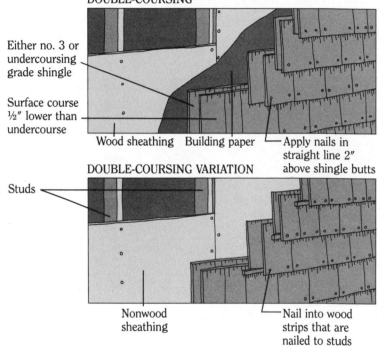

Either no. 3 or undercoursing grade shingle

Surface course ½″ lower than undercourse

Wood sheathing   Building paper   Apply nails in straight line 2″ above shingle butts

DOUBLE-COURSING VARIATION

Studs

Nonwood sheathing

Nail into wood strips that are nailed to studs

the amount of time required to side or roof a house with shingles. Good information on shakes and shingles is available from the Red Cedar Shingle and Handsplit Shake Bureau (Suite 275, 515 116th Avenue NE, Bellevue, WA 98004).

Wood Siding over Foam Sheathing

In their efforts to make buildings more energy efficient, builders have been using insulating foam sheathing rather than plywood or fiberboard sheathing on their homes' exterior walls. It has become apparent that special care must be taken when applying wood sidings over these sheathings. For example, you must provide backing for nails (i.e., you need to provide a wooden surface for the nails to be driven into), usually by furring out the wall with 1-inch boards.

Current recommendations for the installation of wood siding over foam sheathing include:

■ Install lateral bracing for added strength (foam sheathing provides virtually no structural strength).

■ Use a continuous vapor barrier on the interior of the wall.

■ The sheathing and building paper should be allowed to dry thoroughly before the siding is applied. Building paper or an alternative should be used over the sheathing and under the siding.

■ Siding material should be stored under cover until used and should have a moisture content below 16 percent. If the siding gets wet, allow it to dry completely before installation.

■ In the case of beveled siding, thicker patterns are preferable, and a pattern such as a rabbeted bevel that doesn't leave a space behind the board is even more desirable. This reduces the likelihood of cupping and splitting during installation or moisture accumulation over the life of the siding.

■ Use boards of 8 inches or less in width.

■ Use noncorrosive nails that are long enough to penetrate at least 1½ inches into the stud. Ring shanks are recommended for their extra holding power, and their points should be blunted to minimize splitting.

■ Take care not to overdrive nails.

■ All end joints must fall over a stud.

■ Sidings should be prefinished or primed on all sides prior to installation. Brushing or dipping is recommended to assure complete coverage. The finish should be allowed to dry completely before installing the siding, as some solvents will attack the foam sheathing. Regular recoating is necessary for good performance and an attractive appearance.

Finishes

Even decay-resistant woods require care to keep them looking attractive. If the "natural look" isn't important to you, you might decide to paint your siding—this provides superior protection for the wood. If you use air-dried or unseasoned siding, surface preparation may be necessary to keep the moisture in the siding from interfering with the paint film. Check with your paint retailer for product suggestions. High-quality paint is always worth the extra money. Let unseasoned siding air-dry for a month or so before finishing it.

If you don't want to paint, the situation becomes a bit more complicated. At the minimum, a clear water repellent with mildewcide should be applied on the front, back, and all edges of each piece of wood siding before it is installed on the house. It is sometimes possible to order sidings with a factory-applied clear primer over all surfaces of each board. Check with your lumber retailer. If this is possible, remember that you'll still need to saturate the cut ends of the boards with primer.

The lowest-maintenance finish for wood sidings is wood bleach, which speeds the weathering process and lets the wood weather. Your house will end up a silvery gray color. If the original application is uneven or the wood darkens, one more coat may be necessary. This finishing regimen will be more successful on the more stable, naturally rot- and insect-resistant woods such as cedar and redwood. Less stable woods will require more attention to keep them from cupping, splitting, checking, etc., in response to moisture and temperature cycles.

If you want to keep the natural color of the wood, you should talk to building professionals in your area to see what they've had success with. Keeping wood siding looking natural is always a maintenance problem, and although there are a number of products on the market that claim to do the job, they all require periodic reapplication, often in as little as 18 months. We recommend contacting D. L. Anderson and Associates (10650 Highway 152, Suite "U," Maple Grove, MN 55367), the U.S. agent for a Dutch wood finish called Sikkens. It meets all the criteria you'll be looking for in an exterior wood finish. It separates the wood from the elements with a durable and breathable surface barrier and reflects ultraviolet light away from the wood with special pigment particles.

If you're considering treating your siding chemically to make it more decay resistant, take care to find the least-toxic substance that will do the job. *Rodale's New Shelter* has concluded that copper naphthenate, zinc naphthenate, copper-8-quinolinolate, polyphase (3-iodo-2-propynyl butyl carbamate), and TBTO are reasonably safe choices for wood preservatives. Our general feeling about introducing more poisons into our already overtaxed environment is that if it isn't absolutely necessary, don't do it. But if you feel there are no practical alternatives, at least use the substance that will have the least impact on the environment.

Brick and Stone

Brick and stone are virtually maintenance-free under most circumstances, make a very attractive finish, increase the marketability of the house, and give a sense of solidity and value to the home. But laying up brick and stone is a demanding task for a do-it-yourselfer. Further, brick and stone are among the most expensive sidings you can buy, and brick is a relatively porous material, so it doesn't serve as an effective air barrier on the outside of the home.

Stucco

A number of insulation-and-stucco exterior finish systems are available for use over masonry walls. They have attractive features for those building energy-efficient homes. Most can also be used as foundation insulation and over frame walls with the proper sheathing. The best-known of these systems is Dryvit. The only disadvantage of Dryvit is that it is not available to do-it-yourselfers—it is franchised to contractors in order to maintain quality control. But on the plus side, the contractors do an excellent job and offer a generous guarantee against defects. The system consists of an adhesive that bonds the

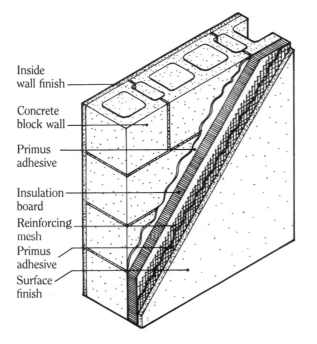

Inside
wall finish

Concrete
block wall

Primus
adhesive

Insulation
board

Reinforcing
mesh

Primus
adhesive

Surface
finish

*Figure 5-9: Dryvit
"outsulation" system.
(Reprinted with permission
from* Fine Homebuilding
*magazine, copyright 1981,
The Taunton Press,
Newtown, Conn.)*

insulation board to the wall, EPS insulation board, fiberglass reinforcing mesh that is embedded in another layer of adhesive, and a finish coat of synthetic plaster. The finish resists cracking, fading, and weather and is available in 21 colors.

Plywood

Plywood siding comes in 4-foot-wide sheets that go up quickly. The material is available in a wide array of patterns and veneers, many of which imitate wood board sidings. The quality of plywood sidings varies—buying a cheap plywood siding is a false economy. If it is not finished or installed properly, plywood siding has been known to delaminate— the layers of wood come apart. Some manufacturers will warrant against delamination for the life of the house, and such a warranty is worth looking for.

Plywood tends to absorb moisture along its edges, so take care to seal all the edges before installing it. For best performance, choose a grade that is free from knots, plugs, patches, and stains. If you plan to paint the siding, purchase a "medium-density overlay" (MDO) plywood, a product with a surface that takes paint well. If you paint other plywood sidings, particularly those with a rough face, the paint may blister, peel, and crack. Plywood siding will check and crack if left to weather naturally.

Most building codes allow plywood siding to be applied directly over studs with no other sheathing material. Although this can result in a fairly flimsy wall, it does have an advantage for the owner-builder building on a tight budget: You can use plywood as an interim siding to satisfy the building department and the lender, then install another siding over it when you have the money. The plywood will then become, in effect, sheathing, serving as a nail base for the new siding and increasing the strength of the wall.

Aluminum, Hardboard, Steel, Vinyl

These sidings are widely available. Our own biases lead us away from them, but you may have different priorities. Many of these products are extremely durable and may be exactly the material you need for your exterior finish. Most are advertised as requiring no maintenance, but they need at least periodic washing, and they may scratch, dent, crack, tear, or eventually fade. Before you decide on a manufactured siding, do careful comparison shopping and, if possible, talk to other homeowners who have had the product you're considering on their homes for a while.

WINDOWS

Windows are a major design feature of your home. How light plays in a room has a dramatic effect on the "feel" of that room. The arrangement of windows also has a large impact on the exterior appearance of your home. Unfortunately, windows let heat out during winter and in during summer. This is why manufacturers have responded to the need for more efficient glazings.

R-Values

The number of layers of glazing in a window and the "emissivity" of the glazing are two of the factors that affect the energy efficiency of a window. A single-glazed window (having just one layer of glass) has an R-value of about 1, extremely low. In windows with multiple layers of glazing, each additional layer adds roughly an additional R-1. A few manufacturers fill the space between the layers with gases heavier than air such as argon, carbon dioxide, or sulfur hexafluoride. These gases provide more insulating value than air.

Double glazing (two layers of glass) is now required by building codes in most areas. In severe climates, the cost of triple glazing may be justified, but the development of high-tech windows such as those with "low-emissivity" (low-E) coatings has led to better alternatives. These coatings applied to the surface of glazings reduce the amount of heat that each layer emits or passes to the next layer. A double-glazed low-E window stops heat more effectively than a traditional triple-glazed window.

A development to watch for is the introduction into the United States of "hard-coat" low-E glass, a product that was pioneered in Europe. The low-E glazings available for the past few years in this country feature a "soft" coating that requires special handling and is very sensitive to moisture. Hard-coat low-E glass can be handled like ordinary window glass. The soft-coat low-E products still outperform the hard-coat, but performance for the hard-coat glass is expected to improve rapidly as the technology matures.

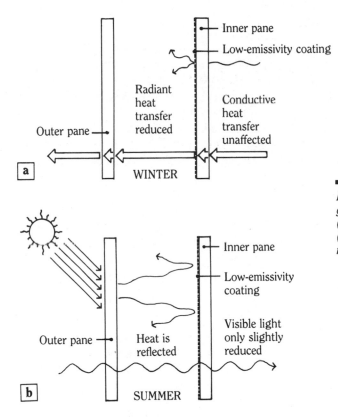

Inner pane

Low-emissivity coating

Radiant heat transfer reduced

Conductive heat transfer unaffected

Outer pane

a | WINTER

Inner pane

Low-emissivity coating

Visible light only slightly reduced

Outer pane

Heat is reflected

b | SUMMER

Figure 5-10: Coated "low-E" glass reduces winter heat loss (a) and summer heat gain (b). (Adapted with permission from Energy Design Update.)

Solar Transmittance

Another measure of window performance is the "solar transmittance" of the glazing. This is the percentage of the total solar energy (the heat and light in sunlight) that passes through a window. A single pane of ordinary glass has a solar transmittance of about 86 percent. An ordinary double-glazed window scores about 71 percent, and triple glazing achieves 59 percent. Low-E coatings reduce solar transmittance. Sungate 100, a double-glazed, low-E unit made by PPG Industries, Inc., has a solar transmittance of 55 percent.

To maximize solar transmittance, "low-iron" glass can be used. By reducing the iron impurities in the glass, manufacturers increase the amount of solar energy that passes through the glazing. Another approach is to use windows made with SunGain, a 4-mil polyester film made by 3M. SunGain is suspended between sheets of glass, serving as another layer of glazing. The film allows more sunlight to pass through than a layer of glass would, it insulates without the weight of another pane of glass, and it transmits less ultraviolet light than glass, thereby reducing fabric fading in the house. One layer of SunGain has a solar transmittance of 93 to 96 percent.

Shading Coefficient

Sometimes you want to keep the sun's heat *out* of your house, especially if you live in the South. The "shading coefficient" of a window indicates how effectively the window blocks solar heat—in a sense, it is the reverse of solar transmittance. Shading coefficients are given on a scale of 0 to 1. A low number indicates that a window is blocking heat effectively.

A window's shading coefficient can be improved by using reflective glass or applying reflective films to the surface of ordinary glass. Unfortunately, many reflective glasses are very shiny in appearance and they limit the amount of visible light that enters the building. The shading coefficients of reflective glazings vary widely, so look for a low-shading coefficient and a high percentage of visible light transmission if you want your windows to look clear but keep the heat out.

Advanced window films (Heat Mirror 55 is an example) can filter out solar heat while still allowing a high transmittance of visible daylight, thus reducing the need for artificial lighting. Electric lights generate heat, which increases the need for air conditioning or other forms of cooling.

Frame Style and Materials

The overall energy efficiency of a window depends on the style of the window and the materials used in its frame, as well as the glazing system. Operable windows that create a tight seal even after repeated use are needed in an energy-efficient home. Casement and awning windows almost always have the lowest air-leakage rates, and unless they clash with the aesthetics of your design, they should be your first choice. At least one manufacturer, Delabro Millwork (2340 South, 3270 West, West Valley City, UT 84119), now makes a sliding window with air-leakage rates that compare favorably with those for awnings and casement windows.

Windows with wood frames usually have higher R-values than those with aluminum, steel, or vinyl frames. Exposed-wood window frames require periodic maintenance, so many manufacturers offer vinyl or aluminum-clad frames to eliminate the need for paint or other finishes. The cladding has the added benefit of increasing the weather-tightness of the unit.

Look for frames with clean, tight-fitting joints. Wood frames should be screwed together rather than nailed together, since this will simplify disassembling the frame for maintenance—something it will need at some point in the life of the window. Many manufacturers publish the R-values, solar transmittances, visible transmittances (the percentage of visible light that comes through the glazing), and shading coefficients for their windows.

We suggest that you use manufactured windows for the operable windows in the house and buy standard insulated glass units from a glass supplier for the fixed windows. If you design the house to use standard sizes, you can save yourself a lot of money. Vertical fixed glazing is not difficult to install, although good detailing and care during the building process is essential to minimize infiltration around the unit.

DOORS

Ideally, the doors in your home welcome you and your friends, keep unwanted visitors out, and hold both air leakage and heat conduction to a minimum.

Doors today are typically made of wood or insulated steel. Wood doors can be purchased prehung with jambs, weatherstripping, and threshold supplied. Unless you know something about hanging doors, we recommend that you purchase a good-quality prehung unit. Hanging doors is fairly skilled work, and an improperly hung door can be a major source of air leakage. The disadvantage of wood doors is that they can warp, split, shrink, and swell in response to changes in temperature and humidity. Always apply a prime coat to an unfinished exterior wood door as soon as it's installed, then finish both sides with a moisture-proof coating.

Steel doors won't change shape or split, although they can dent. Like some wood doors, steel doors are available with foam-insulation cores. This significantly increases the R-value of the door, up to R-15 for some steel doors. A steel door should have a "thermal break" (a strip of insulating material embedded in the perimeter of the door) to reduce heat conduction across the metal. Remember that as much as 80 percent of the heat lost through a door is a result of air leakage, so proper installation is extremely important.

If you plan on any glass doors in your home, we suggest you take a look at the wood-framed atrium-style double doors, since they are usually considerably more weather-proof than sliding glass doors, which tend to allow extensive air leakage. For glazing in your wood or metal entry doors, select double-glazed insulated glass or some of the new high-tech glazing systems we discussed above.

HEATING AND VENTILATING SYSTEMS

One of the major benefits of an energy-efficient home is that it can get by with a radically smaller heating system. But finding these systems is not as simple as going to your local heating contractor and buying a furnace. In a well-insulated and very tight home, the "heating load" (the amount of heat needed) is so minimal that finding a heating system that is small enough can be difficult.

To find out how much heat your home will need, you can do your own heat load calculations or hire someone familiar with energy-efficient homes to do it for you. If you want to do it yourself, a comprehensive workbook is the *Heat Loss Calculation Guide*, available for $8 from the Hydronics Institute, P.O. Box 218, Berkeley Heights, NJ 07922.

One general factor to bear in mind when selecting a heating system is that you may run into problems if the system consumes air from inside the house. The problems may become especially acute if you also have other air-handling devices (such as gas-burning water heaters). These devices will suck air out of the house, thus creating negative air pressure indoors. As a result, outdoor air will try to flow into the house, since air moves from high-pressure regions to low-pressure regions. The outdoor air may come down the chimneys and vent stacks of your air-handling devices, pulling toxic fumes into the house. For example, if outside air comes down your furnace's chimney, it will bring with it the exhaust gases from the furnace. This is a clear health threat. To be on the safe side, we recommend using "direct-vent" heating systems or those that require no combustion air. Direct-vent systems draw air directly from the outdoors into the combustion chamber through an air-inlet pipe and send their combustion waste gases directly to the outside through another pipe.

The Federal Trade Commission requires that a fact sheet be available for each heating system sold in the United States. Be sure to consult the sheets for the systems you are considering. The information on the sheets includes system capacity, model number, estimated annual fuel costs, and a graph that shows the Annual Fuel Utilization Efficiency (AFUE) compared to similar systems of the same size. The AFUE is a calculation of the efficiency of a heating system in normal use over the course of a heating season.

Wood Heat

Wood is one of the few heating fuels that is renewable and that the user can obtain without having to deal with a utility company. A wood heater also gives "focus" to a home, and wood fires generate images of security for most of us. In some parts of the country, firewood is a readily available and economic alternative to fossil fuels.

There is evidence that wood may be a relatively benign fuel environmentally, since by burning it you are speeding up a natural process that would have occurred anyway. A tree rotting on the floor of the forest generates the same amount of carbon dioxide as it would if you burned it completely in your wood stove. There is little question, however, that pollution from wood stove emissions is a serious concern. Problems arise when the wood is not burned completely, when many people in a concentrated area burn wood, and when the air is stagnant.

To shop intelligently for a wood stove, you should have a basic understanding of "combustion efficiency" and "heat-transfer efficiency." Combustion efficiency (the percentage of the potential heat in the wood that is converted into usable heat when the wood is burned) is much more dependent on how a stove is operated than on its design, but the design does have some effect. Some stoves use secondary air inlets in an attempt to assure enough oxygen for complete combustion. While the primary air inlet feeds air to the fire in the stove's combustion chamber, the secondary inlet is located where the air it draws in will cause the smoke rising from the combustion chamber to ignite and burn. This gives you extra heat while reducing pollution. Catalytic combustors are also used in some wood burners. These ceramic devices, shaped like honeycombs, cause wood smoke to burn.

Heat-transfer efficiency is a measure of how effectively a wood stove delivers heat from the fire to you. A stove could have a high combustion efficiency (extracting lots of heat from wood) but a low heat-transfer efficiency (allowing much of this heat to escape unused up the chimney). According to Jay Shelton in his excellent book *Solid Fuels Encyclopedia* (Charlotte, Vt.: Garden Way Publishing, 1983), the features most apt to improve heat-transfer efficiency are, in order of importance:

■ Large exterior surface area relative to the size of the combustion chamber.

■ Keeping hot gases in the stove for as long as possible. In some stoves, the smoke must negotiate a series of baffles before it reaches the chimney.

■ Convection and turbulence both outside and inside the stove. Air should circulate to draw heat from the fire and to bring it into the room. Some stoves use blowers to increase air movement around the stove, with varying degrees of success. If you're considering a fireplace insert (a wood stove designed to fit inside a fireplace), find out if the unit can be operated without its fans on. If it can't, the insert is not much use in a power outage.

■ A stove finish that has a high radiating efficiency. Basically, all colors except metallic colors are good radiators.

Some salespeople and manufacturers claim that cast iron is superior to plate steel as a stove material. The thermal properties of cast iron and steel are virtually identical, so a steel stove of the same thickness holds as much heat as a cast-iron stove. Cast iron is less prone to metal fatigue, which makes it more appropriate for doors where stability is important to ensure a good seal with repeated use. It will crack, however, if it is subjected to thermal shocks or if it is repeatedly heated unevenly. Steel is more flexible, and although it can distort significantly, it's not apt to crack. Cast-iron stoves are often wonderfully decorative, since patterns can be cast directly into the structure of the stove.

How large a stove should you get? The most useful method of sizing stoves is based on their "Btu" outputs per hour. (A Btu is a British thermal unit, approximately as much heat as a wooden match produces.) Most manufacturers indicate how many Btu their stoves can produce. To select a stove, you could calculate the heating load for your home, find out how many pounds of wood you would need to burn per hour to meet this load (all species of wood contain roughly 8,600 potential Btu per pound), and buy a stove that could safely produce that many Btu. Underwriters Laboratories' *Standard for Safety,* UL 103 (ANSIA 131.1) spells out safe burning rates for wood and coal heaters.

The site-built masonry stove is an alternative to the metal wood stove. Masonry stoves can be built in a variety of designs and are known generically in this country as Russian fireplaces and Finnish heaters. If well designed and built, and properly operated, they create almost no pollution and operate at very high overall efficiencies. They are designed to burn small (3 to 4 inches thick), uniform pieces of wood. The enormous mass (literally tons) of these stoves assures a steady, even heat output—they operate on one or two hot, intense fires lasting a couple of hours each day. Once the flames have totally died down, the damper is closed and the masonry radiates heat for many hours. The surface of the masonry never gets hotter than comfortably warm to the touch, so there is less chance of small children or pets hurting themselves than with a metal stove.

There are some potential disadvantages to these heaters, however. Because of their enormous weight, they require their own footings, and because, like any wood heater, they are most effective when located near the center of the house, the house almost has to be designed around them. David Lyle's *The Book of Masonry Stoves* (Andover, Mass.: Brickhouse Press, 1984) is the best introduction to these stoves. It thoroughly covers the subject and includes design and construction details.

Coal Heat

In some areas of the country, coal is used as a heating fuel. In general, the guidelines for combustion and heat-transfer efficiencies are similar to those for wood stoves, but there are important differences you should be aware of before attempting to burn coal as a fuel. Jay Shelton's *Solid Fuels Encyclopedia* is a good source of specific information about types of coal and coal heaters.

We advise against using coal heat. For one thing, coal is dirty, both to handle and to burn. When coal burns, it emits sulfur, which has been linked to acid rain. Coal smoke also contains small amounts of radioactivity. Although coal contains a great deal more heat energy than wood (more than 14,000 Btu per pound for high-quality coal), it also generates a great deal more ash. The coals typically used for home heating have ash contents ranging from about 5 percent to over 15 percent, compared to less than 1 percent for wood. Dealing with this volume of ash is a major consideration in any coal heating system. On the plus side, coal heaters are available equipped with hoppers that will feed coal into the combustion chamber. This makes frequent refuelings unnecessary and lessens the mess of handling the fuel.

Oil Heat

Like coal, oil combustion is a source of sulfur emissions. Another problem is that you may have trouble finding a direct-vented oil furnace or boiler. New oil furnaces and boilers developed over the past few years have greatly increased efficiencies, and advances continue. Unless oil is by far the least expensive fuel in your area, however, we suggest you consider other heating fuels for your new home.

Electric Heat

In many areas of the country, electricity is the most expensive way to heat a home. Since the pollution problems linked to coal-fired power plants and the safety issues associated with nuclear power plants have yet to be resolved, many people are understandably concerned about the environmental impact of electricity. However, there are some real advantages that may make electric heat a viable option in your situation. Electric baseboard resistance heaters and radiant panels are relatively inexpensive to purchase and install, use no combustion air to operate, create no pollution in the home, and allow each room to be equipped with its own thermostat.

Electric heat pumps are considerably more expensive initially, but they can put out two to three times as much heat energy as they consume in electrical energy (that is, for each Btu of electricity, you can get 2 or 3 Btu of heat). "Air-source" heat pumps extract heat from outdoor air, even when the air feels cool to our senses. Unfortunately, as you might expect, these devices tend to be least efficient when it is very cold outdoors. If you live in a frigid climate, a "ground-source" heat pump (one that draws heat from the ground) is probably a better choice.

Electric furnaces and boilers are probably not a good choice for energy-efficient houses, even though they are relatively inexpensive to buy and require no combustion air.

They are harder to control than baseboards or radiant panels, they require the installation of ductwork, and they are less efficient than radiant panels in particular. Furnaces heat the air, which in turn warms us; radiant panels produce heat that warms us directly, without losing as much of its energy in the air.

Gas Heat

Natural gas, where available, is the cleanest and least expensive fuel to use. If the gas-burning system you select is costly to purchase and/or install, however, there may be little real savings. The economic paybacks for high-efficiency heating systems in very energy-efficient homes can stretch nearly to the infinite. Sometimes furnaces that are only moderately efficient make more sense.

Some manufacturers offer furnaces that achieve 80 percent efficiency using conventional technology, but these units are pushed to the limits of their performance. They feature enlarged heat exchangers to extract more heat from the gas fire. This lowers the flue temperatures to about 350°F, which raises the possibility of corrosion in the flue pipe and in the furnace. There are also furnaces with an AFUE of around 85 percent that feature an extra stainless steel heat exchanger. This lowers the temperature of the flue gases to around 200°F, but the stainless steel resists corrosion.

Even more efficient conventional-technology furnaces are available that use additional stainless steel tubing and aluminum-finned heat exchangers. The flue gases from these units are lowered to around 100°F, so that they can be vented through PVC pipe instead of a chimney. This cuts installation costs.

Other furnaces use new technologies to achieve high efficiencies. Amana's Energy Command, for instance, features a "heat transfer module" that combines an electronic igniter, gas burner, and a heat exchanger in one unit and two other heating coils that transfer heat to the house air. Pulse furnaces and boilers work by burning very small amounts of gas and air in short bursts or "pulses." In the Lennox Pulse Furnace, the first pulse forces the combustion gases out a stainless steel pipe, through a secondary heat exchanger, and then through a flue vent to the outdoors. The negative pressure created in the combustion chamber causes the intake valves to open and draw in more gas and air. When the original pulse reaches the end of the stainless steel pipe, part of the flame is reflected back into the combustion chamber, igniting the next pulse without need for another spark. These furnaces have efficiencies as high as 98 percent.

If the heating requirements for your house will be very small, you may have difficulty finding a unit small enough to be practical. It is common in conventional homes to have heating loads of 100,000 Btu per hour or more, while very efficient homes can have loads as small as 10,000 Btu per hour or even less. There are direct-vented, sealed combustion "minifurnaces" that are perfect for this sort of situation, but they may not be readily available in your area. Many are designed to heat perhaps one room in a conventional home, but will often heat an entire energy-efficient house. Minifurnaces are available with a variety of features, including blowers, pilotless ignition, and a variety of Btu ratings. If you have difficulty locating a retailer, try mobile home or recreational vehicle dealers, since they often carry small heating units for use in their products.

Another heating option that is enjoying increasing popularity is combining domestic water heating and space heating in one unit, often a conventional high-efficiency water heater. Hot water from the heater can be sent through pipes in concrete floors or through baseboard radiators to heat the home. Interesting and economical combinations of solar, tankless, and conventional water heating devices are possible, and you are limited only by the heating loads of your home, the hardware available in your area, and your imagination.

Propane Heat

In areas where natural gas is not available, propane is usually an alternative. Most appliances that will run on natural gas will also run on propane with some adjustments, but there are exceptions that you should be aware of. If an appliance is stamped "Natural Gas Only" or "NG Only," don't attempt to adapt it for propane use, since the higher Btu content of the propane could burn out parts designed to be used with natural gas.

Propane is more expensive than natural gas, but in most areas it is a better buy than electricity for home heating. Prices vary widely from one area of the country to another, so you'll have to check with your local suppliers.

Ventilation

In very tight homes, a controllable ventilation system is a must. Proper ventilation is the best insurance against the buildup of harmful pollutants.

It is generally agreed that about 0.5 air changes per hour (ACH) are necessary for healthy indoor air quality. In other words, half of the air in the house should be replaced every hour. The cheapest ventilation technique is to open a couple of windows every once in a while. The difficulty with this is that there is no way of knowing how much is enough. Another possibility is using tight-sealing bathroom and kitchen exhaust fans, perhaps in tandem with humidistats (devices for measuring humidity) and automatic vent openers. Since much of the moisture and odors generated in a home are generated in bathrooms and the kitchen, thoroughly ventilating them may suffice for the entire house, especially in a relatively mild climate.

Another way to ventilate tight homes is with air-to-air heat exchangers. These devices recapture most of the heat from the outgoing stale air and use it to preheat the incoming fresh air. They do this by blowing the two streams of air through a heat-exchanger core, without letting them physically mix with each other. Thus the heat from the outgoing air—but none of the substances in this air—is passed to the incoming air. Air-to-air heat exchangers are available in whole-house sizes and smaller room-size models.

There is some question about when air-to-air heat exchangers make economic sense. The expense can be considerable for the whole-house models and the payback slow, depending on your situation. Costs range from $600 to $1,600, installed. Before you make any hard decisions, check with your local energy office or builders and designers that are familiar with the paybacks for your area. In very general terms, it seems that an air-to-air heat exchanger can pay for itself in a reasonable amount of time if you live in a very cold climate in a very tight house, and if you heat with an expensive fuel, such as electricity. Otherwise, it probably makes more sense to use a cheaper ventilation method.

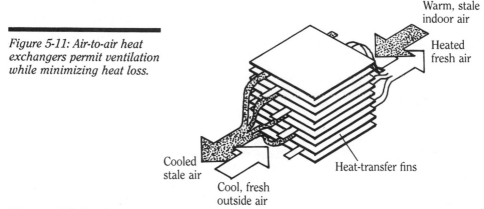

Figure 5-11: Air-to-air heat exchangers permit ventilation while minimizing heat loss.

Warm, stale indoor air

Heated fresh air

Heat-transfer fins

Cooled stale air

Cool, fresh outside air

Formaldehyde

One of the most common indoor pollutants is formaldehyde, a colorless gas. Formaldehyde is highly toxic, yet it is commonly found in drapes, furniture, tobacco and wood smoke, cosmetics, permanent-press clothing, towels, hair sprays, grocery bags, newsprint, soap, household disinfectants, and toothpaste! Drinking less than 1 ounce of liquid formaldehyde can kill you, and the human body is highly sensitive to formaldehyde fumes—many people's eyes will water at 1 ppm (parts per million) in the air.

To minimize the amount of formaldehyde in your new house, you can avoid the use of the plywoods and particleboards that are the worst offenders. This isn't as difficult as it sounds. The U.S. Department of Housing and Urban Development (HUD) has issued standards that limit the amount of formaldehyde that particleboard and plywood may emit. (The standards are intended for mobile homes, but they can be applied to other structures as well.) The products that meet the standards carry a stamp to that effect. The HUD standards limit particleboard emissions in mobile homes to 0.3 ppm and paneling emissions to 0.2 ppm.

Plywoods and other products made with urea-formaldehyde glue generally emit ten times as much formaldehyde as products using phenol-formaldehyde adhesive. So another tack is to look for phenol-formaldehyde products. These include most exterior-grade plywoods, waferboards, and composition boards, as well as phenolic particleboards. The phenol-formaldehyde resins are more expensive than urea-formaldehyde, but in our view they are well worth the extra expense if your family's health is at stake.

There are some ways you can cut down the emission levels from products already in the home. Most paints and sealers that are good vapor barriers are also formaldehyde barriers, and particleboard covered on all sides with plastic laminate is unlikely to be a problem. The highest levels of emissions occur when the heat is first turned on in a new house. After about a month, the level will drop dramatically, often by a factor of ten. Extra ventilation will be required for at least a year, and after that you should maintain at least 0.3 to 0.5 ACH. Sweden requires that new public buildings be aired out for six months by using outside air at an unusually high ventilation rate, and many experts suggest similar programs for new homes.

CHAPTER 6

Land, Money, and Red Tape

In this chapter, we will present information about subjects that are not directly related to the design and construction of your home but that nevertheless can have a profound impact on your project.

BUILDING SITES

Land costs represent an increasingly large portion of the cost of building a house—roughly 24 percent, on average. In many areas, the percentage is significantly higher, due at least in part to government bureaucracy, land-use controls, zoning ordinances, and environmental protection policies. Many of these regulations are critical to assure quality of life in the areas affected, but unfortunately they often inflate land costs. You should study the regulations in the community where you plan to build, since they can have an important effect on your plans.

According to Frank Gray, former city planning director for Petulama, California, and Boulder, Colorado, there are basically two zoning systems used in the United States. The first, referred to as a "cascade" zone system, permits all uses allowed in less restrictive zones, plus the uses specified for the zone in question. For example, single-family homes are usually considered the most desirable use for a particular site, followed by multifamily residences, commercial buildings, and industrial buildings. If you wanted to build a single-family home in an area zoned for multifamily homes, you could. But you would not be allowed to build a duplex or triplex in an area zoned for single-family structures. The second type of zoning is the "specific-district" system in which only those uses specifically defined in that particular zone are permitted. Uses permitted in less restrictive zones are not allowed. Thus, you could not build a multifamily unit in a "specific-district" area unless such units are specifically permitted there, regardless of whether they are permitted in less restrictive districts.

For both cascade and specific-district zoning, it is important that you check not only the lot you are interested in, but also the zoning on the surrounding properties. It would be disheartening to build your dream house and then learn that a waste dump was going to be created next door.

Some communities have "solar access" ordinances that protect homeowners' access to the sun (so your neighbors can't erect any structures that would shade your house), but such ordinances tend to be rare, so check out the situation carefully. Height restrictions might be of particular importance if you plan to build a solar home, or if there are attractive views you want to preserve.

Ask the building department for a list of all the agencies that will review your building permit. Give particular attention to flood control and seismic considerations, since either can make a lot virtually unbuildable. It would be a shame to purchase a lovely piece of land with a picturesque stream running across it only to discover that you can't build on that land because it's in a floodway.

In addition to restrictions imposed by local governments, many subdivisions have their own covenants that you must conform to if you wish to build there. Covenants typically cover aesthetic questions such as the type and style of permissible structures, the exterior finish materials that may be used, and what kinds of domestic animals are allowed. Many subdivisions have active architectural committees that review house plans before they may be built. If you have your heart set on building a large redwood dome home, and the lot you're considering is in an architecturally conservative subdivision, better check to see if there is an active architectural committee before you buy the land.

Finding a Lot

If you're new to the area, drive around and familiarize yourself with what's available in buildable lots. Good real estate agents can be worth their weight in gold, since they should be well acquainted with restrictions that might apply to the piece of land you are considering. Keep in mind, however, that real estate agents work for the seller, and the responsibility to research potential problems with a site is yours.

When you find a lot that appeals to you, get specific in your research. Find out what the asking price is and how flexible the terms (if any) are. It is often possible to buy land for a little money down and have the seller carry the note for a specified amount of time at attractive rates of interest. There is little risk for the seller, since if you default on your payments, he or she will get the property back and you will lose whatever you have in it. If you plan to build out of pocket, having the seller carry a note might be an ideal arrangement, since it frees your cash for materials and other house expenses.

If you plan to borrow from a bank to build and don't want to pay cash for the land, find out if the owner is willing to "subordinate" to the construction loan. This means the owner takes a secondary position to the construction lender, so that if you default, the lender gets paid before the owner. This is the minimum a lender will require; more often he or she will require that you own the land free and clear. In that case, you may be able to use the first installment of the construction loan to complete the land purchase.

If the price of the land is within your budget, it's time to find out if the property meets other criteria. If the lot seems like a particularly good deal, we would urge you to bring a healthy dose of skepticism to the negotiating table. There are any number of things that can add significantly to the development costs on a piece of property, and even though the selling price may seem attractive, by the time the lot is buildable, you may have spent much more than you had expected to.

The following is a list of things to consider. If you're concerned about losing the lot while you look into these matters, it's possible to write a contract naming any or all of the following considerations as contingencies. In some areas, obtaining a building permit is itself contingent on meeting these criteria, and if this is the case in your building jurisdiction, making the land contract contingent on obtaining a building permit might serve as a blanket safeguard.

Property Location

The location of the lot, the pattern of growth in the surrounding area, and how the lot and your proposed construction project fit that pattern will all affect the land's value. A single-family dwelling sandwiched between shopping centers will be less marketable than one in an attractive residential neighborhood.

The condition and arrangements for maintenance of streets and roads also affect property values, and in many areas you must be on a maintained road to get a building permit. In many parts of the country, it is possible for a lot to be in a subdivision and still not have legal access because maintenance requirements have not been met. Local governments can require that all buildable lots be on maintained roads.

Federal flood insurance regulations have been tightened considerably. In many areas, it is impossible to get a building permit or mortgage for a property in a floodway or floodplain.

Soils

Ask for a copy of any soils report that has been prepared about the lot. If none is available, check with local building officials to see if they will require a soils report before they issue a building permit. They probably will make such a requirement if other homes in the immediate area have had trouble with expansive or weak soil, a high water table, or other soil problems. Soils reports are expensive ($400 or more), so if there is an existing report, or if you can talk the seller into paying for one, you will save money. If there is a problem with the soil, the building department will almost certainly require that the foundation be engineered, increasing your costs.

Check to see if the lot you're considering is in an area where bedrock is close to the surface. If your dream house is earth-sheltered or includes a full basement, you may have to blast away the rock before you build—not the most practical or cost-effective way to excavate a site. A high water table can also make below-grade living space impractical.

Water Supply

If there is a water line adjacent to the lot, find out what the fee is to hook up to it. If there isn't a water line, look into the possibility of obtaining a well permit. The local driller can tell you how deep the wells are in the immediate area, which will give you a rough sense of the cost.

Building departments usually require some evidence of a water supply before they'll issue a building permit, and usually a well permit or letter of commitment from the local water district suffices. A lender, however, often requires that the property already have a producing well (assuming there is no other water supply available), and most lenders specify how much water the well must provide. If your well produces less water than the lender specifies, you may be required to install a cistern or make other arrangements to assure a constant supply of water.

Sewage Disposal

Building departments and lenders require adequate sewage disposal. If no sewer line is available, you can usually build a conventional septic system. If, however, soil conditions make such a system impermissible, you should examine a number of alternative systems, including composting toilets, incinerating toilets, specially engineered septic systems, and "gray water" recycling systems. Many of these options make good sense, but lenders and building and health officials may hesitate to approve them. We would encourage you to explore the possibilities, but leave yourself enough time to plough through the red tape you're likely to encounter.

Septic systems generally must be kept a minimum of 100 feet from any well or water course, and that distance can increase to 200 feet if there is fractured bedrock or a high water table on the lot. Give some thought to how you can arrange the house, septic system, and well on the lot without endangering your own or a neighbor's well or pond. It's always preferable to locate the septic system downhill from the house, but if this arrangement is impossible, "lift stations" (devices designed to pump sewage uphill) can be installed.

The test used to determine whether the soil is appropriate for a septic system is called a "percolation" or "perc" test (a procedure that measures the rate at which soil absorbs water). A perc test isn't difficult to perform, and there may be places on the property that test out better than others. It is often in your interest to do your own perc tests before buying the land. Rent or borrow a post-hole digger and dig three holes 30 to 36 inches deep in the area that you plan to use as a leach field (the portion of septic system that allows water from the main septic tank to leach out into the soil). Fill these holes with water and keep them full until they are saturated and the rate at which the water drops in them has stabilized. Then time how long it takes the water in the hole to drop an inch. The result is known as a "perc rate." In some areas, a perc rate of 1 inch in anywhere from 5 minutes to an hour is acceptable. In other areas, the rate can be slower or faster, as long as the size of the leach field is adjusted accordingly.

Utilities

Call the local utilities and find out what it costs to get any services you will want (electricity, gas, phone service). One owner-builder bought a beautiful mountain lot and discovered some months later that running electric lines to the site was going to cost over $4,000! Find out, also, how long it will take from the time you order the service until it is actually installed. One woman was shocked to find that it would be at least two months before the local utility could connect the temporary electricity at her rural lot. In order to start work on her home, she had to buy a portable electrical generator.

A phone may seem unnecessary on a building site—until the first time one of the tradespeople working at the site makes an unauthorized decision because he or she didn't want to drive to a phone and talk things over with you. Even if you plan to be on site most of the time, a phone can be invaluable to call for supplies and information. It can also be a godsend in case of an emergency. Contact your local phone company early if you plan to have a site phone.

It is a courtesy to your subcontractors and neighbors to rent a portable toilet in all but the most remote areas. For the rental fee, usually around $60 a month, most companies clean and empty the toilet every week.

Demographic and Economic Indicators

A home's value is affected by the condition of the local community. Factors you should investigate include population growth or decline and the reasons for it; employment/unemployment figures; and how many homes in the area are vacant. You should also consider how convenient your specific building site is to local schools, churches, shopping facilities, recreation areas, and public transportation.

Resale considerations dictate that you keep the house within 15 to 25 percent of the value of the average home in the neighborhood. You will also find the house easier to sell if it is conventional looking, since it will presumably appeal to a broad spectrum of consumers. Of course, if you are reasonably sure that you will not want to sell the house for the foreseeable future, this becomes less of a concern.

Trees, Terrain, Views

Mature trees can add as much as 25 percent to appraised land value. Although they can also add to construction costs, since you will have to build around them, the aesthetics and increased property value are well worth the trouble.

Hillside or otherwise difficult lots can be the most valuable in the long run, if they are developed thoughtfully. A major advantage to south-sloping lots is that they are oriented toward the sun (which is always in the southern sky: southeast in the morning, southwest in the afternoon). This makes solar heating easier.

Very steep lots can be a problem, however. Development costs may be prohibitive, and slippage of the soil can make a property unsuitable for building. If there's a question in your mind, request a geologic report before buying. One indication that there may have been slippage in the past is the phenomenon of "pistol butt" trees—trees with dramatically bent trunks. If all or most of the older trees on a hillside have a pronounced pistol butt at the base of the trunk, it is an indication that the ground shifted, and the trees compensated for the shift.

Miscellaneous

Make sure there is a recent and accurate survey of the property. One owner-builder fell in love with a piece of mountain property that seemed to be an unusually good deal. He asked the owner for a copy of the survey, and was directed to some stakes that had been driven into the ground on the property. When he insisted on another survey, he discovered that the stakes were inaccurate: The existing driveway was on the adjacent property. If he had built where he'd planned, part of his house would have been on the neighbor's lot. Eventually, the sellers bought enough of the adjacent lots to include the driveway and the planned house site, but it took a full year to straighten everything out.

Ask if there is a title insurance policy on the property to assure that the title is free from any liens (claims by creditors). The policy should be updated before you buy the lot. Also find out if there are any easements or rights-of-way across the land.

It will probably take some time to do all the necessary research concerning the lot you want to buy. In the meantime, don't put down more than 10 percent earnest money, and be sure it's held in escrow if you're working with a real estate agent. It's a good idea to check with your attorney before making any final decisions.

FINANCING

Historically, financing was one of the largest obstacles the aspiring owner-builder faced. Lenders are a notoriously conservative lot, and they generally prefer to deal with established, professional builders. A number of lenders are starting to show signs of receptivity to owner-builders, however. Some actually prefer lending to owner-builders, since they find that owner-builders are more committed to their construction projects than professional builders for whom the work is "just a job."

Finding a Receptive Lender

If you have a good working relationship with a specific bank that makes construction loans, that's the place to start looking for money. If you already have a track record of successfully managing loans and other relationships with this institution, you should find a receptive ear when you outline your project. If your banker is unwilling to make the loan, however, ask him or her for names of loan officers at other banks who might be more receptive and ask if he or she would be willing to write you a letter of introduction.

Other owner-builders are often a good source of names of receptive lenders. Ask around. If there is an owner-builder school in your area, it probably has a list of lenders they work with, or the school could at least put you in touch with some of their former students.

Many lenders prefer not to lend in rural areas. Lenders are mainly concerned with the marketability of the property, and building in a rural or remote area narrows the prospective resale market considerably. There are other lenders, fortunately, who have made an informal specialty of lending for projects on remote or difficult sites. If you're building in a rural or relatively remote area, leave yourself more time to research your financing.

Convincing a Lender

Many lenders have had the experience of being left with a half-finished house to sell and would prefer not to repeat the experience. Since they are in business to lend money, not build houses, lenders are not equipped to act as technical consultants during a building project, and they understandably don't want to spend much time helping owner-builders through crises. Once they have made the loan to you, however, they have a very real vested interest in having the house completed in a timely and efficient manner, and most will do whatever they can to expedite the process.

In some ways, it's useful to treat your transaction with a lender as if you were applying for a small-business loan. From the lender's point of view, you are an unknown entity: You have no track record at the activity you want to borrow the money for. You must impress the lender with your commitment to the project, the thoroughness of your preparation, and your ability to repay the debt.

Prepare in detail for your meetings with the lender. At your first meeting, you'll want to have:

■ A complete set of house plans and specifications, assembled with enough detail that it is easy to tell what the house will look like, how it will all go together, and what materials will be used to build it. If an engineered foundation is necessary, get the engineering done and the plans stamped. In many areas, the building officials require that all foundation designs be approved by a structural engineer.

■ Detailed cost estimates. Construction lenders will check the accuracy of your estimates, and they may require that you change your figures to conform to theirs, if there is a conflict. Since your house will probably be custom-designed and custom-built, it is entirely possible that your estimate will be more accurate than the lender's. One woman encountered this difficulty with her construction lender. When the house was built, it turned out that her estimate was correct—almost to the board—and the lender's was way off. In order to get the loan, however, she had to modify her figures to more closely match the lender's.

When preparing your cost estimates, make some preliminary contact with subcontractors. Quotes from subs and suppliers will carry more weight than cost estimates that you work up yourself.

■ Proof of ownership of your land or a subordination agreement with the seller. Clear title to your land and the title insurance to confirm it is always a huge plus in the eyes of the lender. Many lenders will work only with owner-builders who own their lots outright. The advantage to the lender is that ownership gives you more equity in the project and thus increases your incentive to complete the project. (It might be useful to have an "owner equity" column on your cost-estimate sheets. Since bankers deal in dollars, figure out the actual dollar amount that you will save by doing all or part of the work yourself and show that figure in the "owner equity" column.)

■ Evidence of a producing well or other water supply and an approved septic permit or other evidence of sanitation arrangements. Note that the septic system needn't be installed at this stage, only designed.

■ A demonstrable knowledge of the building process, utility requirements, zoning restrictions, subdivision covenants, and local building codes. Managerial, planning, and scheduling expertise are all useful on construction jobs, so if you can demonstrate that you have a history of these kinds of experiences in other areas of your life (if it's a part of your regular job, for instance), it will also strengthen your case. Some lenders recognize the diploma from an owner-builder school as an indication that you are serious about gathering as much information as you can about the project, and that you have ready access to the kinds of help you will need as things progress.

■ Financial information:

1. Proof of current income. Some lenders call your employer to verify employment, while others ask for income tax returns for the two or three previous years. If you are self-employed, you will have to provide tax returns.
2. Evidence of a good credit rating.
3. A personal financial statement of what you own and what you owe. The lender will be concerned not only about your current debt-to-income ratio, but also the stability of your job history. The lender is considering committing to a long-term obligation, possibly as long as 30 years, and he or she will be a good deal more comfortable if you seem likely to be gainfully employed 10 years down the road.

Because the lender may be concerned about your ability to complete the parts of the project you plan to do yourself, he or she may require you to qualify for a larger loan than you think you need. The lender's purpose will be to ensure that, if necessary, you will be able to hire workers to do jobs you planned to do yourself. Actually, this is a good idea. Since you pay interest only on the money you draw from your construction loan, taking out a larger loan will cost you only a little extra for the front-end fees that are based on the loan amount. This may prove to be cheap insurance against future problems.

If a lender is very skeptical of your ability to build a house, consider offering a guarantee of completion. This could take a number of forms. Jim Patton, former director of the Durango Owner Builder Center, suggests putting together a package that includes evidence of completion of a housebuilding course at an owner-builder school; an agreement with a general contractor that if anything should happen to you, he or she would step in and finish the house; and insurance against acts of God, so that the house would be completed even if the wildly unexpected should occur. Also included would be a requirement that you contract for consultations with either your owner-builder school or a qualified contractor. Once construction begins, you would be required to submit to the lender and the contractor periodic statements of progress and expenses to date. If you've fallen behind the agreed-upon schedule, or if the budget is getting out of hand, the contractor would have the option of stepping in and finishing the job.

The Mechanics of Conventional Financing

There are two parts to an owner-builder's financing package: the short-term financing or construction loan, which is used to cover construction costs, and the long-term mortgage loan, which you get after construction is complete. Typically, you would use the mortgage loan to pay off the construction loan. You would then pay off the mortgage loan over a period of time, often 30 years, in monthly installments.

In order to obtain a construction loan, you must first secure a commitment from a lender for a mortgage loan. The mortgage lender agrees to pay off the construction loan and provide permanent financing once the house is completed, assuming that (1) the house is completed according to the plans and specifications, and (2) your financial situation hasn't changed substantially in the time it took to build the house. This agreement is also called a "permanent commitment" or a "permanent take-out."

Mortgages

To get your relationship with the mortgage lender off on the right foot, be straightforward about your background and financial history. There are many things, some of which are out of your control, that will affect your chances of getting financing from this lender. It is important that you talk to a person who has the authority to make at least a preliminary decision, and that you have a good personal interview with him or her.

There are so many types of mortgages that it would require another book to cover them all. In most of the new types of mortgages, the homebuyer shares the risk of fluctuating interest rates with the lender. One word of advice: Avoid any adjustable-rate mortgage (ARM) that is "negatively amortized." In such an arrangement, the principal amount of the loan actually increases for a period of time, which means you could make a number of payments on your mortgage and end up owing more than you did at the beginning of the loan! Other snares you should avoid are loans with interest rates that can float freely and loans that put no limit on how much your payment can increase each year and over the life of the loan.

Shop carefully for a mortgage and don't be embarrassed to ask for clarification if there are points you don't understand. The complexity of the current money market demands that consumers educate themselves thoroughly about the financing vehicles they're considering. Don't assume that all lenders are out to scalp you, however. A loan that ends in default is ultimately not any more attractive to mortgage insurers than it is to consumers. The competition for money is so intense these days, that lenders can ill afford to develop a reputation for taking advantage of their customers.

The type of mortgage you get will depend on your financial situation, the type of house you are building, the mortgage plans available in your area, and your personal preferences. The mortgage lender will tell you how much money you qualify for on the basis of the payment you can make each month. Lenders want to make sure that your monthly payment does not exceed a predetermined percentage of your income, often between 25 and 30 percent, after other obligations have been considered. One recent development that holds promise for owner-builders is a push to allow individuals who are building exceptionally energy-efficient homes to use a larger percentage of their income for making mortgage payments. The rationale is that lower utility bills will leave more money available for mortgage payments. Check with your local lenders to see if they offer such a program.

In spite of how thoroughly you research your options, some things are out of your control. Interest rates are likely to remain high, limiting the supply of money available for home mortgages. There is even the slim possibility that there won't be affordable money available when it comes time to retire your construction loan. When you qualify for your mortgage commitment, you qualify at the prevailing rate at that time. When your house is done, 6, 9, or 12 months down the road, rates could rise—hypothetically, at least—to a level where you no longer qualify for the mortgage.

In spite of such grim possibilities, we are not trying to discourage you from tackling the project. Rather, we urge you take the best precautions you can. Your best insurance is to plan your project with meticulous care, and if possible, plan a home that will cost you less than what you can qualify for. The comforting historical reality is that when extreme situations develop, the individuals and institutions involved usually manage to come up with a solution that leaves the owner with the house. Lenders don't want your house; they just want their money back, and there isn't a single instance we've been involved with where the owner-builder lost his or her house.

Construction Loans

Having secured a mortgage commitment, the next step is to convince a construction lender to give you a construction loan to build the house. Some institutions do both construction and mortgage lending, but more often you will have to shop elsewhere for construction money.

Traditionally, construction loans for single-family residences have been made for a maximum of six months. But lenders accustomed to working with owner-builders will often make the loan for as long as a year. It's a good idea to make arrangements ahead of time for extensions in case you need them. Once construction starts, you'll have your hands full.

You will probably have to work harder to convince the construction lender of your abilities than you did the mortgage lender. The letter of commitment that you get from a mortgage company contains a clause that states that the company will provide permanent financing "upon satisfactory completion" of the home. In essence, this means that the mortgage lender is under no obligation to provide financing unless you complete the project in a timely and acceptable manner. If you fail this test, the construction lender could be left holding the bag—so this lender will want strong assurances from you before extending a loan to you.

As you look for a construction lender, familiarize yourself with the prevailing rates in your area so that you'll know how good a deal various lenders are offering. Most construction loans are set up so that you pay interest only on the money that you actually use. Always check your loan agreement and make sure. Paying interest from the beginning on the total amount of a construction loan can break your budget in a hurry.

Once you've been approved for a construction loan, the lender will set up a checking account for you. All the funds for the construction of your house should be paid through this account. If you have some cash that you want to put into the project before you use the bank's money, it's best to arrange for the construction loan first, deposit your funds in the construction checking account, and draw on them until they're used up, at which time you would begin to draw on your construction loan. If there is money left in the account at the end of the project, you will have paid a bit more than necessary in "front-end" costs when you took out the loan—but this should be far cheaper than using the entire loan and paying interest on all of it.

Many lenders do not want you to spend any money on your project until a construction checking account is set up, and then they want you to make all payments through this account. If you start spending money before the account is opened, they may refuse to extend a loan to you. The reason has to do with liens. Anyone who does work or provides supplies on a construction job has the right to file a lien against the property if he or she isn't paid. This means that, if the claim is legitimate, this person is entitled to payment before the property can be bought, sold, or refinanced. The lien would have to be paid before you could retire the construction loan and get a mortgage loan. Naturally, construction lenders want to be as sure as possible that no liens are filed. If all the money you use has been funneled through the construction account, the lender has reasonable assurances that everyone has been paid, because the canceled checks provide evidence. To reinforce this evidence, you should also collect lien waivers from everyone you write checks to. The easiest way to do this is to have a lien waiver printed on the back of the checks, so when tradespeople endorse the checks, they indicate that they waive the right to file liens. Lien waiver forms are also available at office supply stores.

Associated Expenses

Both the construction and mortgage lenders will charge for a variety of things besides the interest on the loans. Often these costs must be paid in certified funds when loan arrangements are completed, so it's important that you be aware of what they are. The following is an example of the costs you might expect to encounter:

■ Title insurance policy. Construction lenders require title insurance to assure them that you own the building lot free of liens or other encumbrances. You pay for the lender's policy, and it's also a good idea to buy one for yourself because the lender's policy covers only the lender's liabilities, not yours. If a dispute should arise over the title, the lender would be protected but you could lose your investment. The fee for the second policy is usually nominal. When you finish the house and get your mortgage loan, the mortgage lender will require an update of the policy. Again, get a policy for yourself at this time.

■ Front-end or origination fee. This covers administration and other costs the lender incurs in putting the loan together. It is usually 1 "point" or 1 percent of the loan value.

■ Appraisal. Both the construction lender and the mortgage lender will require appraisals based on the plans and specifications, and then the mortgage lender's appraiser will walk through the house after completion to assure that it was constructed according to those plans and specs. Sometimes construction lenders have their own appraiser and include the cost of the appraisal in the origination fee, but more often you'll pay for it separately.

■ Credit report. The construction lender will require a credit report separate from the one required by the mortgage company. The mortgage company requires one at the time of application and an update when the house is complete to assure that your financial status hasn't changed in the time it took to build the house.

- Recording fees. About $3 per page in many areas; this covers the cost of recording the documentation for the loans.
- Taxes and insurance. Your property taxes should be fully paid when you take out the construction loan, and you will be required to buy a "builder's risk" insurance policy to cover the project until it is complete, at which time the policy will be converted to a homeowner's policy. Your property taxes and homeowner's insurance are included each month in your mortgage loan.
- Survey. A survey is done to assure that the proposed house will be on the lot you own, that it won't encroach on any easements, and that the setbacks are correct. The survey is often not required until the foundation is poured or even later, but you would be wise to have it done before beginning construction.
- Construction loan interest. Some lenders have you pay the construction loan interest on a monthly basis as it accrues, while others let you wait until the project is complete. Often it is incorporated in your mortgage payment.
- Other costs specific to your situation or to the lender. This might include a real estate tax service, which authorizes the title company to send the annual property tax statement to the mortgage company so that they can include the appropriate amount in your monthly payment. (Title companies are organizations that search real estate titles to ensure that they are free of liens and that all easements are identified.)
- "Junk fees." Lenders may charge for nearly anything the consumer will agree to pay. You may find you are being charged for document preparation, underwriting, etc. Remember the Golden Rule: "He who has the gold makes the rules." But don't hesitate to ask for a clear explanation of all charges.

Cutting Costs

Naturally, you will benefit if you can keep the size of your construction and mortgage loans as small as possible. Here are some suggestions:

- The most obvious way to cut financing costs is to reduce the size of the house. While this may not seem desirable, we think it merits consideration. You could build a small, expandable house designed with expansion in mind, so that you can enlarge it later (see chapter 3). Starting small and expanding later has several advantages. If you do most of the work yourself, you may have enough equity in the property to get a home improvement or equity loan to continue the work. You will also have a place to live during the rest of construction.
- If you're designing the house yourself, find out what the standard sizes are for the materials you'll be using. For example, framing lumber comes in 2-foot increments (studs are normally 8 feet, 10 feet, or 12 feet long, for example), and sheet goods (plywood, insulative sheathing, etc.) come in 4-foot increments (4 feet, 8 feet, 12 feet, etc.). If you design your house with these dimensions in mind, you'll have much less waste, so costs for materials will be lower.
- Some hardy owner-builders live in a temporary shelter on their land while they build

their home. They can then apply the savings from rent or mortgage payments toward materials. The disadvantage is that it can be stressful to live primitively while working hard. It may also be illegal to live in a trailer, teepee, tent, or other shelter in some areas, so check the local regulations before you proceed with this plan.

Less Conventional Financing Options

If you can't work out financing with a conventional lender—or if you don't want to— you should consider various alternatives:

■ Ask family and friends. There may be someone among them with enough cash and faith in your abilities to make the construction loan—or at least to advance you enough to significantly lower the amount you'll need to borrow from a construction lender. If you take this route, keep the transaction businesslike. Sign a note so that your friend or relative will have recourse, should problems arise, and offer to pay a rate that is competitive with the rates they could get from other short-term investments.

■ If you currently own a home, consider selling it and using the profits to finance your new home. You might be able to build the new home out of pocket or at least significantly reduce the amount you'll need to borrow. One possible scenario, if you have the time and enough equity, is to sell your present home, put a small down payment on a piece of land and let the owner carry the note, then use the rest of the profit from the house sale to build.

■ If you've seen any kit homes that appeal to you (see chapter 4), see if the manufacturers offer construction financing or if they are willing to help you find financing. Some kit manufacturers finance only the actual construction of the house, not site improvements. Check into this and be sure that you are aware of all the costs involved.

■ Look into local community bond issue funds or federal agencies such as the Veterans Administration (VA) or the Farmers Home Administration (FmHA). If you have the patience and stamina to wade through their paperwork, and if you're qualified, you can sometimes save significant amounts. The VA requires very little or sometimes no money down on some loans. FmHA has a self-help housing program that might be worth exploring.

■ If the chips are really down, boldness may be in order. One owner-builder started his project with cash he had and then ran ads in the newspaper offering good rates on an investment secured by real estate (the home he was building). He got so many responses that people were actually bidding against each other. He ended up with a very competitive arrangement and is now living in his house. Of course, there's no guarantee that such a risky scheme would work for you.

Estimating Building Costs

Complete and correct cost estimates are critical to the success of any building project, but they are particularly important to the owner-builder, since you can't recoup your losses on the next job. Professional estimators use several methods to arrive at an accurate estimate. As a preliminary "ballpark" figure, they will cite the average square-foot costs of comparable homes in the area. Since the chances are that you are considering a customized house design that has never been built before, the only way this method would be useful is if you're buying from one of the few companies that are manufacturing customized homes designed on a module. In this case, you could get an idea from the manufacturer how much it has cost in the past to build the modules you choose.

By far the most most accurate method of cost estimating is called quantity surveying. This involves a "takeoff," or detailed itemization, of all the materials that will go into the house and a "takeoff" of the labor that will be necessary to make all those materials into a house. Overhead and profit are then added to the sum of these figures. For the owner-builder, overhead and profit are the equity you will have in the project when it's finished. While no estimate is infallible, and indeed sometimes cost overruns seem to be the rule rather than the exception on construction jobs, absolute thoroughness and attention to detail are your greatest allies as a novice estimator. Literally try to list the prices of all the materials you will use in constructing the house, and then—using advice from tradespeople and other owner-builders—carefully calculate how much you'll have to spend to hire out each construction task that you do not intend to handle yourself.

The healthy enthusiasm and optimism that owner-builders bring to their projects sometimes puts them at a disadvantage when they get down to the nitty-gritty of figuring out how much their dream house is going to cost them. Emotional attachment to desirable but nonessential items can spell disaster for a tightly budgeted construction project. We've watched owner-builders convince themselves, in the face of professional advice and lots of evidence to the contrary, that their budgets were sufficient to justify purchasing the hot tub or custom-built cabinets early in the project. You should make contingency plans that allow for the frills only if all the work up to that point has gone according to plan and budget.

The estimating process, while time-consuming and often tedious, is one of the most effective and comprehensive planning tools at your disposal. It will make it possible to shop for materials with increased confidence, and the expertise you gain will make it easier to convince a lender that you're competent to build your home—you'll know in detail what all the parts and tasks are. For this reason, even if you're building out of pocket, we urge you to take the same care with your estimate that you would if you were presenting it to a lender.

Most office supply stores carry estimating forms, and construction lenders often have forms they use. Anything that helps you stay organized and keeps you from forgetting various items will serve the purpose. Ask around in your area and see what the contractors use. Table 6-1 presents a sample form.

Recognizing that doing a takeoff on your home is a long process, you should tackle it in short blocks of time rather than trying to sit down for 6 or 8 hours at a stretch. Your mind will get fatigued in a couple of hours, and you'll be much more apt to make mistakes and get discouraged.

A careful examination of your plans and specifications, looking for inconsistencies and discrepancies, should be one of your first tasks. Specifications always take precedence over the drawings, so pay particular attention to them. Measure everything exactly as it shows on the drawings—do not approximate, average, or round off your figures. If there is a

Table 6-1 ESTIMATING FORM

Job: _____

_____

_____

_____

Materials: $ _____

Labor: $ _____

Other: $ _____

Tax: $ _____

= Total $ _____

MATERIALS

Material	Dimensions	Quantity	Unit of Measure	Cost per Unit ($)	Total Cost ($)
	X X				
	X X				
	X X				
	X X				
	X X				

LABOR AND EQUIPMENT

Item Description	Owner	Skilled Help	Other	Duration	Quantity	Unit of Measure	Cost per Unit ($)	Total Cost ($)

discrepancy between the dimensions you get when measuring the drawings and the dimensions that are written on the drawings, go with the written dimensions until you can check with the person who drew the plans. (If you drew the plans yourself, figure out where you made a mistake and correct it.) Similarly, if there are discrepancies between the drawings and the specifications, accept the specifications until you can resolve the discrepancies. Measure everything that you see in the drawings, and make sure it is all reflected in the specifications. Never leave something out because it seems insignificant. As a contractor we know says, "Take it all off."

List materials and tasks in the order they will occur and keep the separate tasks separate to avoid compounding errors. For instance, as you start, figure the excavating work for the foundation separately from forming and pouring the footings; figure forming and pouring the footings separately from forming and pouring the foundation walls; and so on. This makes mistakes easier to isolate and will make comparing the cost of different materials and techniques simpler. If, for example, you decided to use concrete block for the foundation rather than poured concrete, you would only have to refigure the foundation walls.

Segmenting the project this way also makes it more psychologically manageable. It is very easy to lose your perspective when dealing with hundreds of details and large amounts of money. Pouring a concrete footing, framing a floor, or tiling a bathroom floor are much easier and less overwhelming to cope with than building a whole house.

Never round off your results until you finish the calculations for the task you're estimating. Say you're calculating the amount of concrete you'll need to pour your footings. The footings are 8 inches deep by 16 inches wide, by a total of 130 linear feet. Eight inches is ⅔ of a foot, and 16 inches is 1⅓ feet. If we multiply this out, we get 115.584 (.667 × 1.333 × 130) cubic feet of concrete. We need to convert this to cubic yards (concrete is measured in cubic yards), so we divide 115.584 by 27 (the number of cubic feet in a cubic yard), and we get 4.281 cubic yards. Having arrived at this result, you can now round up by 5 percent or so to allow for waste or irregularities in the footing trench: Call it 4.5 cubic yards. But never round off your numbers before reaching this point. If you'd gotten lazy and rounded your ⅔ of a foot up to 1 foot, your calculations would look like this: 1 × 1.333 × 130 = 173.29 cubic feet. 173.29 ÷ 27 = 6.418 cubic yards. This is almost 2 cubic yards more than the correct answer. This is a gross example, but it illustrates the point. If you round things off before you get to the end of the operation, you're apt to end up with a lot more material than you want or need.

As you proceed with your estimates, write down all your thinking as you go along. If you're doing it right, someone else should be able to sit down with your figures and easily trace the path that got you to the bottom line. Use a printing calculator for your math and attach the calculator tapes to the estimate sheets so that you can refer to them. As you figure parts of the job, mark them off on the plans so you don't figure something twice.

Most Common Estimating Mistakes

Following are some suggestions to help you avoid mistakes that are frequently made in preparing construction estimates.

■ Many people fail to familiarize themselves with site conditions. There is a wealth of information to be had at your building site. Familiarity with your land and the surroundings can save you large amounts of time, money, and hassle.

As you walk around the property, visualize how things will happen. Are there rocks, outbuildings, trees, or debris that must be removed? Do you have the time, skills, and equipment to take care of the job or will you have to pay someone else? Is there room for large trucks to clear the utility wires? Is there a road or a place for a road large enough to accommodate big trucks? Is the ground hard enough to support heavy trucks and equipment? Think about how workers and materials will get in and out of the site in all kinds of weather. Remember that if materials can't be delivered directly to the building site, you must move them or pay someone else to move them from where they're dropped to where the house will be.

How deep are the wells in the area? Is a conventional septic system possible? Does the property include large trees that must be removed? (If trees must be removed, be sure to leave 3 or 4 feet of stump so that a tractor driver can catch them with his bucket and push them over. It can mean the difference between $40 or so per tree, and as much as $200 per tree if the trees are cut down to small stumps that have to be pulled from the ground.)

■ Excavation work is very difficult to estimate; this is probably an area where you will need professional help.

Once the excavation is done, where are you going to put all the dirt? Will it have to be hauled off? Needless to say, hauling dirt around costs money, and it's quite possible there will be no room to store a big pile of earth on a small lot. Will blasting be necessary? A license is required in most areas to handle explosives, so this is not a do-it-yourself operation. Will you be building on solid, undisturbed earth or a landfill? Building codes require that you compact the dirt in a landfill or sink caissons (concrete piers) down to solid earth.

■ Underestimating "incidental" costs can cause serious problems for novice home-builders. These include such items as tool rentals and purchases; loan costs; insurance; permit fees; septic, structural, and/or soils engineering fees; road building and maintenance; storage sheds for tools and materials; a site telephone; a portable toilet; sales taxes; keeping sharp saw blades on the site; all those unplanned trips to the lumberyard; and construction cleanup.

■ If you have the money or can qualify for it, estimate for the highest possible quality materials and labor. It is always possible to downgrade the specifications if that becomes necessary, but you're in trouble if you have cost overruns at the beginning of the job, and your estimate is for a bare-bones house. Lenders are also more comfortable if they know that you can cut back without damaging the marketability of the house.

■ To repeat: The most common mistake we see owner-builders make in their estimates is not taking enough time to become thoroughly familiar with their project. The result is that they leave something out of the estimate. One person forgot the footing drainage system; another didn't allow for any door hardware; and another didn't investigate the subsurface conditions of his lot thoroughly enough and had to blast through bedrock to run the water line from the well to the house. Throughout your project, you will be making up for inexperience by arming yourself with an intimate knowledge of your particular house. Spend as long as it takes to know that building inside and out and to get a clear idea of your options at each stage of the process.

■ Always develop contingency plans and leave yourself a cushion to take care of the items that are inevitably overlooked. Once you've done as thorough an estimate as you're capable of, relax and enjoy your project. Of all the people we've watched build homes as first-time owner-builders, not one has failed to complete the job. The ones that ran into trouble were, to a person, the ones who didn't think it was necessary to plan carefully and took a "design as you go" approach to the project. But even all of them eventually finished and now live in the homes they built.

CODES AND PERMITS

Nearly any contractor you talk to will have at least one story to tell about an unreasonable building inspector. But in fact, the building inspector is often an invaluable ally and source of information to the novice builder. It is in your interest to cultivate a positive, cooperative relationship with him or her early in the project. Keep clearly in mind that building officials have the authority to shut down your project if they think there is a code problem with what you're doing.

The Model Codes

In the United States, there are three building codes that most local building departments use as the models for their own codes. The Building Officials and Code Administrators International, Inc. (BOCA) publishes the Basic Building Code, used by some communities east of the Mississippi. The Southern Building Code Congress International, Inc. (SBCC) publishes the Standard Building Code, used mainly in the South. The International Conference of Building Officials (ICBO) publishes the Uniform Building Code (UBC). We will discuss the UBC mainly, since it is the most widely adopted, but we urge you to find out what code is in force in your area.

Some local building departments adopt model codes in their entirety, and sometimes they make slight changes. Other communities write their own codes, usually based on one of the model codes. Whatever its origin, the local code determines what and how you can build, so you should get a copy (including the provisions for mechanical, plumbing, and electrical work) and study it. You will likely be impressed by the detail that is presented in the code book. The complete UBC covers all kinds of buildings besides residences, which is largely useless information for the owner-builder. Fortunately, the ICBO publishes an abridged version of the UBC that deals only with residential construction. It is considerably cheaper and less formidable than the complete UBC. Many building departments sell copies over the counter, and it is also available in some bookstores. If you have trouble finding it in your area, write to the ICBO, 5360 South Workman Mill Road, Whittier, CA 90601 or call (213) 699-0541 and ask for "Dwelling Construction under the Uniform Building Code." If another model code is used in your area, check to see if they have a similar residential version of their code.

In the words of the UBC, the purpose of the codes is to "provide minimum standards to safeguard life or limb, health, property and public welfare by regulating and controlling the design, construction, quality of materials, use and occupancy, location and maintenance of all buildings and structures within this jurisdiction and certain equipment specifically regulated herein." Your house will almost certainly be built to higher standards than the local code requires, but in any event, you must be sure to meet all code requirements.

The codes primarily cover conventional construction techniques. If you are considering using unconventional materials or building methods in your home, you can anticipate some resistance from building officials. They will want assurance that your materials and methods will perform as well as the ones listed in the code. Section 105 of the UBC states that "the provisions of this code are not intended to prevent the use of any material or method of construction not specifically prescribed by this code, provided any alternate has been approved and its use authorized by the building official." As long as you can convince the building officials in your town that your plans are safe and effective, you can probably get your technique or material approved.

If you feel the building department is being unnecessarily conservative, Section 204 of the UBC also states "In order to determine the suitability of alternate materials and methods of construction and to provide for reasonable interpretations of this code, there shall be and is hereby created a Board of Appeals consisting of members who are qualified by experience and training to pass upon matters pertaining to building construction." This gives you recourse if you can't get any satisfaction from your building inspector. According to the code, the Board of Appeals consists of other building professionals who may be more sympathetic and better able to assess your situation than the building department.

One owner-builder, Bruce Gandrud, is a classic example of how one can work successfully with local government. Gandrud wanted to build an adobe home in Boulder, Colorado. It would be the first adobe home in the city. Boulder building officials were predictably skeptical about the suitability of adobe to the local climate. But Gandrud discovered that the state of New Mexico has written an excellent amendment to the UBC that covers adobe and other earthen structures. This gave him a precedent to present to the Boulder officials. Earth homes are common in many parts of the world with much wetter climates than Boulder's, and investigation revealed that there were numerous ways of designing and building adobe structures for such climates. Gandrud had to make some design compromises to win approval from the Boulder officials, but his adobe home is now finished, and he and his family are happy with the results.

Permits

Regardless of whether you're building an unconventional house or a standard structure, you will have to get a building permit before you begin construction. The procedure for obtaining a permit varies from one building department to the next, but to give you an idea of what's involved, we'll describe what one building department requires.

Permits in the sample area are good for 18 months, and you may not suspend or abandon work for a period exceeding 180 days, or you will have to renew your permit in order to begin work again. You must also renew your permit if the house isn't completed in the 18-month period.

The building department requires you to submit the following in order to obtain a building permit:

■ An application form, available at the building department.

■ A soils report, if there is any evidence of expansive soils, high water table, or other problems.

■ A copy of your deed recorded with the county clerk.

■ Evidence of a water supply.

■ Evidence of arrangements for sanitation.

■ Proof of legal access to your site (if the property does not front on a county road).

■ Two sets of construction drawings and specifications that illustrate all the proposed work in enough detail to clearly show compliance with building codes and zoning regulations. The set of plans must include:

1. A plot plan (a plan of the property with the house on it).
2. A foundation plan.
3. Elevations (front, rear, and side exterior views of the house).
4. A floor plan of each level.
5. Cross section(s) of the house.
6. Details of critical elements that clearly show sizes, materials, connections, and construction, if they apply to your project including (a) window details, if not shown on floor plans, (b) fireplaces and chimneys, (c) roof truss designs with stress analysis or name of manufacturer, (d) roof hips and valleys, decks, stairways, and any other applicable elements.

If necessary, the building official may request other information to confirm the structural adequacy of the structure.

Inspections

As work progresses, you will be required to arrange for inspections by a building official as predetermined stages of the project are completed. You are required to have the inspection record card displayed in a conspicuous spot on the site at all times, where the inspector can conveniently make entries on it, and you must also have a copy of the construction drawings on the premises.

It is a good idea, and in many areas a requirement, to give the inspector at least one working day's advanced notice of your request for inspection. You must provide the inspector with easy access to the work to be inspected.

The following are the inspections that are typically required by building departments. (Check to see what inspections are required by your local department.)

Rough Inspections

■ The temporary electric construction pole must be inspected after all wiring installations have been made as required in the National Electrical Code, 1981 edition, Article 305, entitled "Temporary Wiring."

■ Trenches, footings, pads, and caissons must be inspected after the trenches are excavated and the forms erected, prior to pouring the concrete.

■ Foundation walls and/or grade beams (continuous footings reinforced to act as beams) must be inspected after all forms are erected, prior to pouring the concrete.

■ Damp-proofing of footings, foundations, and/or grade beams for basement walls must be inspected prior to backfilling.

■ Any electrical work that is to be located under a concrete slab within the building must be inspected prior to backfilling and/or the pouring of the slab. If the meter housing is mounted on a pole, the building official must inspect between the pole and the house prior to backfilling.

■ Any plumbing work that is to be located under a concrete slab within the building must be inspected prior to backfilling and/or the pouring of the concrete slab.

■ Any gas piping that is to be located under a concrete slab within the building must be inspected prior to backfilling.

■ Rough electrical work must be inspected prior to being covered and concealed. Inspection will be made of all walls, ceilings, floors, the service equipment panel, and the service entrance.

■ Rough plumbing work must be inspected prior to being concealed in walls, ceilings, and floors. Waste lines, vents, and supply lines will be included. Whether or not it is required in your area, we recommend that you pressure test supply systems and drain-waste-vent systems (DWV) before closing the walls.

■ Rough gas piping must be inspected prior to concealing in walls, ceilings, and floors.

■ Rough heating and ventilation work must be inspected prior to concealing in walls, ceilings, and floors.

■ Rough framing must be inspected after the roof, all framing, fire blocking, and bracing are in place, and all pipes, chimneys, and vents are complete. Note: The roof covering (shingles or other) must be on in order to pass this inspection.

■ Insulation in walls, ceiling, and floors must be inspected prior to concealing in walls, ceilings, and floors.

■ Lath and/or wallboard must be inspected prior to plastering or taping, and after all nails and/or screws are installed.

Many of these rough inspections can be done simultaneously, e.g., rough electrical, rough plumbing, rough gas piping, rough heating and ventilation, and rough framing. Inspectors appreciate not having to run back and forth to your site four or five times to make inspections they could make in one visit.

Final Inspections

Final inspections are usually made after work on the building and site is completed and the home is ready for occupancy. If all is in order, you will be issued a certificate of occupancy and may live in the house with the blessings of your building department. Mortgage lenders require a certificate of occupancy before they make the permanent loan on the property. In many areas, it is possible to get a temporary certificate of occupancy before the house is entirely complete, if the building official determines that "no substantial hazard would result from occupancy."

The following inspections are often required to get a permanent certificate of occupancy. (Check to see what final inspections are required in your area.)

■ An inspection of the distance between buildings and property lines, as well as rights-of-way.

■ Final grading is inspected to see that there is positive drainage away from the house—the inspector is not concerned with whether the landscaping has been done. Your lender, however, may require that the landscaping be completed, since this affects the marketability of the property.

■ The frame is inspected to demonstrate that all rooms and areas are complete in every respect.

■ Electrical work is inspected. All fixtures must be installed and operational, and the service entrance and service equipment panel must be complete.

■ All plumbing fixtures must be installed and functional.

■ An inspection of gas piping, which entails an air-pressure test of not less than 10 pounds per square inch. The pressure must be maintained for at least 15 minutes.

■ Heating and ventilation work must be complete in every respect and in compliance with the local code.

■ Window glazing and insulation is inspected. Double glazing is now required by code in many areas, and the "Federal Glazing Act" requires safety glass in any doors intended for human passage, tub or shower enclosures, and windows located within 12 inches of a door and below the line of the top of the door.

Potential Problems

If you start work without a permit, you are liable for an "investigation fee," in addition to the cost of obtaining a permit. The investigation fee is often equal to the cost of the permit, so the result is that your permit fee is doubled—and the inspector will keep a close eye on you for the rest of your project. If any work has been covered so that it can't be inspected or is not in compliance with the code, the inspector can require you to tear it out and redo it. If concrete has been poured without an inspection, you could conceivably have to redo the foundation. It is your responsibility to know what is required of you and comply. Some jurisdictions, for instance, require separate permits for various parts of the job, such as plumbing and electrical work, and some even require that homeowners planning to do parts of the job take tests to prove minimum competency.

Any time an inspector is unhappy with the work you are doing, he or she has the authority to stop work on your project (sometimes called "red tagging" the job). If you think the stop work order is unjustified, you have the right to appeal it, but you may not proceed with work until the appeal is settled. Since this can upset all your careful scheduling and cost extra interest on the construction loan, it is always in your interest to establish and maintain an open, cooperative relationship with the building inspector.

APPENDIX A

PERT Charts

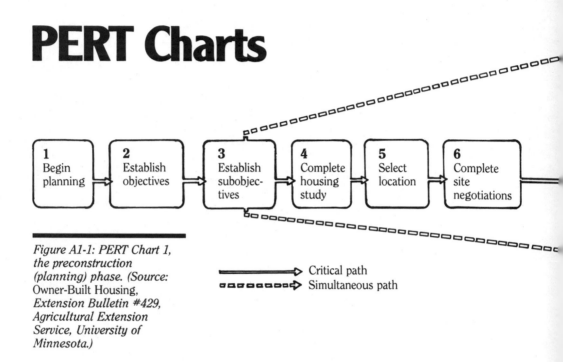

Figure A1-1: PERT Chart 1, the preconstruction (planning) phase. (Source: Owner-Built Housing, Extension Bulletin #429, Agricultural Extension Service, University of Minnesota.)

Getting people, materials, equipment, and money organized into a smoothly flowing construction sequence is critical to the success of your project. As a means to that end, here are PERT (Performance Evaluation and Review Techniques) diagrams that show the "critical path" and "simultaneous path" for a typical owner-built house. The critical path indicates events that must occur in the order presented, since some tasks cannot begin until others have been completed. The simultaneous path represents tasks that can occur at the same time.

Although your job will probably differ from these PERT charts in some ways, you should be able to use the charts as a guide. You may want to make your own diagrams. They can be as detailed as you think appropriate. For instance, they could include the amount of time you expect each step to take, the individuals who will be involved, and how much money you expect to have spent by that point. Including such information will help you tell quickly if you're over budget or time.

172

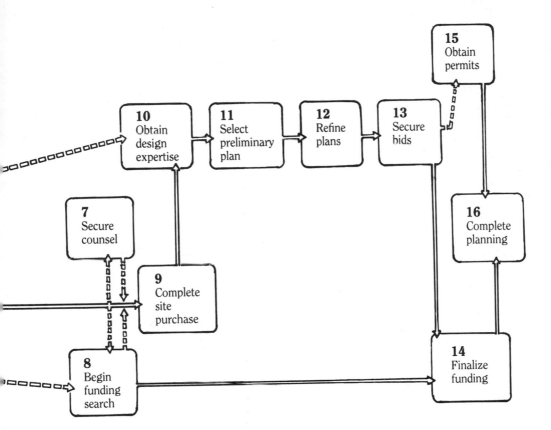

Planning Phase

1. Determine what land and construction costs are in your area, what functions you want the house to perform, and what your level of involvement in the project will be.

2. Establish specific design objectives for the project. Establish time and budget constraints at this point.

3. Start your land search. Explore alternative materials and building techniques. Make preliminary contact with designers and lenders.

4. Complete your housing study by taking classes, reading, visiting construction sites, contacting local building officials, and talking with owner-builders, materials suppliers, and tradespeople. Consider working with a builder or owner-builder to gain some hands-on experience.

5. Establish the exact location of the house on the site where you want to build. Check with local building, zoning, and planning officials before making a final decision.

6. Complete land negotiations. Agree on a price, perhaps pay earnest money, and sign a preliminary sales agreement.

7. Secure legal counsel before finalizing the deal, if only to have your lawyer read all the paperwork and make sure it is in order.

8. Start your funding search now. Talk to lenders, friends, relatives, and governmental agencies that might be able to help.

9. Complete land purchase.

10. Obtain design expertise from an architect, architectural designer, builder who also designs, plans service, or other source.

11. Decide on a preliminary design.

12. Once you're happy with the design, draw (or have a professional draw) the final plans for the house. Show the plans to your building official and lender for approval, then complete the materials list and specifications and your estimate of construction costs.

13. Secure bids from materials suppliers. If you'll be hiring out any of the work, solicit bids from subcontractors.

14. Finalize your funding arrangements. Establish a construction schedule and check with your subs and suppliers to be sure they can meet it.

15. Obtain a building permit. (Sometimes the subs will obtain the permits necessary for their parts of the job.)

16. Do anything else necessary to complete the planning process. This usually involves staying in touch with the building department, lender, subs, and suppliers, and resolving any problems that arise.

Figure A1-2: PERT Chart 2, the building phase. (Source: Owner-Built Housing, Extension Bulletin #429, *Agricultural Extension Service, University of Minnesota.)*

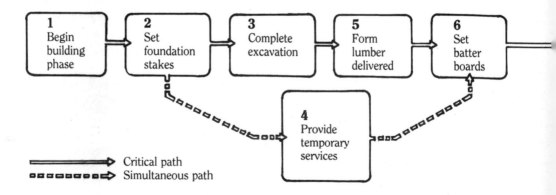

Building Phase

1. Make preliminary arrangements for temporary services such as electricity, toilet, telephone, and water at the construction site and erect a construction shack to store tools and materials.

2. Prepare for the excavation: Set the stakes that define the outline of the house and remove trees, brush, or outbuildings. Provide space to store soil that will be used for backfilling. Arrange to protect trees or other plants that you want to shield during construction.

3. Complete the house excavation.

4. Install and hook up the temporary electrical service. If you'll need water on the site for masonry work or other reasons, finalize these arrangements now.

5. Take delivery of form lumber for the footings. If you are building an all-weather wood foundation, the gravel for the footings should be delivered at this time.

6. Place "batter boards" (horizontal boards nailed to the stakes) to define the perimeter of the foundation.

7. Dig and form the footings. Check the height, level, and squareness of the footings and make sure the ground is firm. Codes generally require that footings be poured on undisturbed or mechanically compacted soil.

8. In most areas, footings must be inspected and approved by a local building inspector before concrete can be poured.

9. Pour the footings.

10. Take delivery of foundation block or, if you will be pouring a concrete foundation, the form boards.

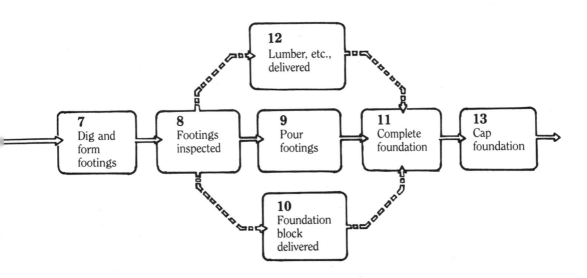

Figure A1-2: PERT Chart 2, continued.

⟹ Critical path
⟹ Simultaneous path

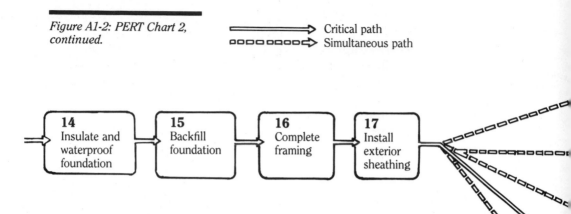

| 14 Insulate and waterproof foundation | 15 Backfill foundation | 16 Complete framing | 17 Install exterior sheathing |

11. Place the forms for a poured concrete foundation, then have them inspected by the building inspector. Order the concrete, pour the walls, and strip the forms.

12. Take delivery of lumber and other supplies for framing and enclosing the house.

13. Install sill plates, floor joists, and subflooring.

14. Waterproof and insulate the foundation. Install drain tile.

15. Backfill around the foundation, compacting every foot or so to guard against future settling.

16. Frame the house. Construct exterior and interior walls, roof, and stairways.

17. Install sheathing on exterior walls and roof.

18. Install rough plumbing. Get all water, drainage, and vent lines ready for inspection.

19. Install heating and cooling systems. Check for proper routing of ductwork and clearances for chimney.

20. Install wiring for electrical system, phone, cable TV, security systems, doorbell, etc. Check that outlets, switches, fixtures, thermostats, doorbells, etc., are in the right places.

21. Install the roof, paying attention to flashing details at any penetrations through the finish roofing.

22. Install windows. Check for proper operation and fastening. Note: Some windows are designed to be installed after the siding is up (see step 24).

23. Have framing, plumbing, wiring, and roofing inspected. This is usually the most important inspection, so take care to have everything ready.

24. Complete the exterior of the house. Install siding, doors, and trim. Check doors for proper swing and tight closure.

25. Make preparations for concrete slabs, walks, and drives. Check for correct depth and compaction of excavation, and check for adequate reinforcement of the concrete in slabs, walks, and drives.

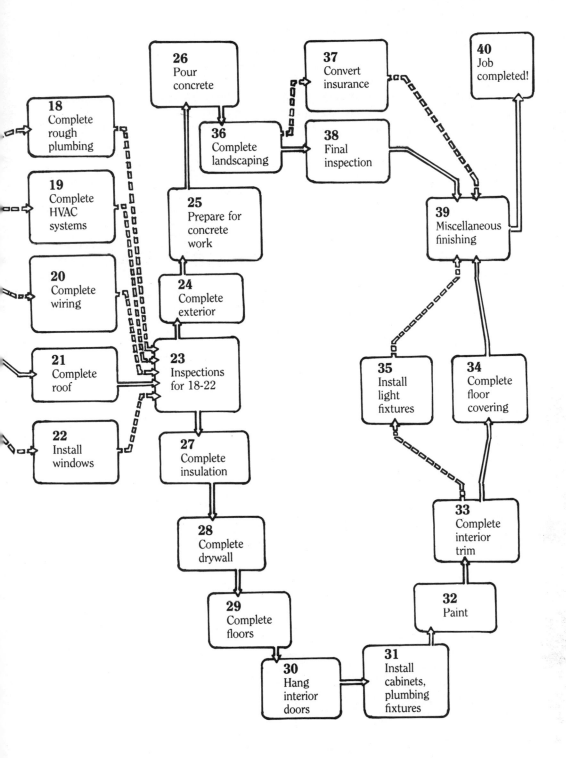

26. Complete the pouring of concrete areas. Check that concrete surfaces have the desired finishes and that the pitch of slabs is correct.

27. Install insulation and air/vapor barriers in walls, roof, floors.

28. Hang, tape, and sand drywall surfaces.

29. Install underlayment over the subflooring. The finish treads may be added to stairways, but protect them from wear during the rest of the job.

30. Install interior doors. Check for proper swing and closure and make sure adequate clearance has been left for final finish floors.

31. Install cabinets and plumbing fixtures. Check for proper operation.

32. Paint interior of home.

33. Complete the interior trim. Paint or stain trim before installing it. Check for clean joints throughout the house.

34. Complete the finish floors. Check carpeting, tile, linoleum, hardwood floors, etc., for proper colors, patterns, finishing, etc., and resolve irregularities with subcontractors.

35. Install lighting fixtures. Check to be sure that all the lights work from the proper switches.

36. Landscape. Check for proper drainage before landscaping begins and check for correct plantings and other site work once it's complete.

37. Convert insurance coverage to new home.

38. Have the final inspection performed. Receive certificate of occupancy. Retire the construction loan and obtain permanent loan (mortgage).

39. Finish any miscellaneous odds and ends.

40. Job completed. Have a party—you deserve it!

Resources for Owner-Builders

Following is a list of organizations around the United States and Canada that offer instruction in various types and phases of housebuilding to prospective owner-builders. Contact those that interest you for more information and ask in your area for other sources of instruction. Many state and local energy offices, for instance, provide building programs and energy conservation programs. Some businesses also teach classes related to their products.

Owner-Builder Schools

Building Resources
121 Tremont St.
Hartford, CT 06105
(203) 233-5165

Colorado Owner Builder Center
P.O. Box 12061
Boulder, CO 80303
(303) 449-6126

Cornerstones Energy Group, Inc.
54 Cumberland St.
Brunswick, ME 04011
(207) 729-6701

Durango Owner Builder Center
502 Ludwig Dr.
Bayfield, CO 81122
(303) 884-9021

Earth Resource Technology
Blue Mountain Rd.
Wilseyville, CA 95257
(209) 293-4924

Earthwood Building School
RR 1, Box 105
West Chazy, NY 12992
(518) 493-7744

Eastfield Village
Box 145 RD
East Nassau, NY 12062
(518) 766-2422

Ember, A Building School
P.O. Box 208
Taos, NM 87571
(505) 776-8913

Georgia Solar Coalition
1083 Austin Ave. NE
P.O. Box 556
Atlanta, GA 30307
(404) 525-7657

Heartwood Owner-Builder School
Johnson Rd.
Washington, MA 01235
(413) 623-6677

Homebuilding Institute
2424 N. Cicero Ave.
Chicago, IL 60639
(312) 745-3901

Miami-Dade Community College
Owner Builder Center
11011 S.W. 104th St.
Miami, FL 33176
(305) 596-4113

Michigan Owner Builder Center
1505 E. 11 Mile Rd.
Royal Oak, MI 48067
(313) 545-7033

Minnesota Trailbound School
of Log Building
3544½ Grand Ave.
Minneapolis, MN 55607
(612) 822-5955

Natural Spaces, Inc.
Rt. 3, Box 105 D
North Branch, MN 55056
(612) 674-4292

Northwest Owner Builder Center
13240 Northup Way
Suite 27
Bellevue, WA 98005

The Owner Builder Center
1516 5th St.
Berkeley, CA 94710
(415) 526-9222

Owner Builder Center of New York
160 W. 34th St.
New York, NY 10001
(212) 736-4909

Owner Builder Center at Sacramento
6 Colbey Ct.
Sacramento, CA 95825
(916) 961-2453

Riverbend Timber Framing, Inc.
P.O. Box 269012
East U.S. 223
Blissfield, MI 49228
(517) 486-4044

School of Log Building
P.O. Box 1238
Prince George, BC V2L 4V3
Canada
(604) 563-8738

Shelter Institute
38 Center St.
Bath, ME 04530
(207) 442-7938

Southwest Solaradobe School
P.O. Box 7460
Albuquerque, NM 87194
(505) 842-0342

Timbercraft Homes
85 Martin Rd.
Port Townsend, WA 98368
(206) 385-3051

Traditional Craftsmen
Log Building School
6464 Burleson Rd.
Oneida, NY 13421
(315) 363-2028
or
Two Bar Seven Ranch
Tie Siding, WY 82084
(307) 742-6072

University of California
Santa Barbara Extension
Owner Builder Program
University of California
Santa Barbara, CA 93106
(805) 961-3695

Urban Shelter Associates, Inc.
1252 S. Shelby St.
Louisville, KY 40203
(502) 636-3663

Pat Wolfe Log Building School
RR 1
McDonald's Corner, ON K0G 1M0
Canada
(613) 278-2009

Yestermorrow, Inc.
P.O. Box 344
Warren, VT 05674
(802) 496-5545

Other Informational Sources

Arcosanti
HC 74 Box 4136
Mayer, AZ 86333
(602) 632-7135

Brick House Publishing Co.
3 Main St.
Andover, MA 01810
(617) 475-9568

Cosanti
6433 Doubletree Rd.
Scottsdale, AZ 85253
(602) 948-6145

Davis Energy Group
123 "C" St.
Davis, CA 95616
(916) 753-1100

Delabro Millwork
2340 S, 3270 W
West Valley City, UT 84119
(801) 972-2383

D. L. Anderson and Associates
10650 Highway 152
Suite "U"
Maple Grove, MN 55367
(612) 424-3344

Dryvit System, Inc.
One Energy Way
P.O. Box 1014
West Warwick, RI 02893
(800) 556-7752

GeoTech Systems Corp.
100 Power Ct.
Sterling, VA 22170
(703) 893-1310

Log Home Guide Information Center
Exit 447
Interstate 40
Hartford, TN 37753
(800) 345-5647

McIBS, Inc.
8000 Maryland
Suite 450
St. Louis, MO 63105
(314) 721-8401

Poly Plastic and Design Corp.
1920 E. Pleasant St.
P.O. Box 299
Springfield, OH 45501
(513) 323-4625

Raven Industries, Inc.
P.O. Box 1007
205 E. 6th St.
Sioux Falls, SD 57117
(605) 336-2750

Red Cedar Shingle
and Handsplit Shake Bureau
515 116th Ave. NE
Suite 275
Bellevue, WA 98004
(206) 453-1323

Red Rocks Community College
12600 W. 6th Ave.
Golden, CO 80401
(303) 988-6160

Shakertown Corp.
P.O. Box 400
Winlock, WA 98596
(206) 785-3501

Sto-Cote Products, Inc.
Drawer 310
Richmond, IL 60071
(815) 675-2358

Storey Communications, Inc.
Garden Way Publishing
Schoolhouse Rd.
Pownal, VT 05261
(802) 823-5811

Timber Framers Guild of North America
RR 1, Box 207
Alstead, NH 03602
(603) 835-6840

Viking Penguin, Inc.
40 W. 23rd St.
New York, NY 10010
(212) 337-5200

A 50-page Resource Directory listing contact information for products and services useful to owner-builders is available for $4 from the Colorado Owner-Builder Center, P.O. Box 12061, Boulder, CO 80303.

Index

Italic numbers indicate photos or illustrations; boldface numbers indicate graphs or tables.

Rodale Press, Inc., publishes RODALE'S PRACTICAL HOMEOWNER™, the home
improvement magazine for people who want to create a safe, efficient,
and healthy home. For information on how to order your subscription,
write to RODALE'S PRACTICAL HOMEOWNER™, Emmaus, PA 18049.